PHYSICS AND CHEMISTRY OF
MOLECULAR
ASSEMBLIES

PHYSICS AND CHEMISTRY OF
MOLECULAR ASSEMBLIES

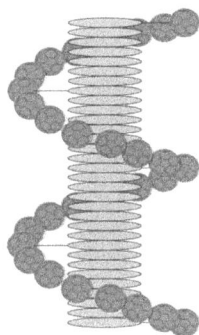

Kazuchika Ohta
Shinshu University, Japan

World Scientific

NEW JERSEY · LONDON · SINGAPORE · BEIJING · SHANGHAI · HONG KONG · TAIPEI · CHENNAI · TOKYO

Published by

World Scientific Publishing Co. Pte. Ltd.

5 Toh Tuck Link, Singapore 596224

USA office: 27 Warren Street, Suite 401-402, Hackensack, NJ 07601

UK office: 57 Shelton Street, Covent Garden, London WC2H 9HE

British Library Cataloguing-in-Publication Data
A catalogue record for this book is available from the British Library.

ISBN 978-981-121-578-0

For any available supplementary material, please visit
https://www.worldscientific.com/worldscibooks/10.1142/11703#t=suppl

Typeset by Stallion Press
Email: enquiries@stallionpress.com

Preface

You may comprehend unconsciously that all natural matter can be categorized into only three states of solid, liquid and vapor. By reading this book, you will change this traditional notion of matter into a new modern recognition, and finally, you will agree that the states of matter in nature is diverse. The knowledge acquired in this book will be very useful for a wide range of research and technology for liquid crystal displays, drug efficacy in pharmaceuticals and many others. You will profoundly understand the real practical basic knowledge of thermodynamics and X-ray structural analysis, which are the foundation of all research

When a large number of molecules assemble to form the aggregation states they display many functions which have never been seen in one molecule. Moreover, even if they consist of the same molecules, each of the aggregation states gives completely different physical properties: density, hardness, optical and electrical properties, drug efficacy, etc In order to pursue the functionality of materials, it is indispensable to understand molecular assemblies of crystals, liquid crystals, glasses, colloids, etc

In this book, I have focused on the basics of crystal polymorphism and liquid crystal polymorphism, and have described a novel concept of matter, which was not recognized until now, and novel X-ray structure analysis methods for many two-dimensional and one-dimensional phases. In order to profoundly understand the concepts of this book and actively utilize the novel methods presented here, it is very important for the readers to resolve the exercises of each chapter themselves. The answers to these exercises will be provided by the author as a sequel to this book in the near future.

Kazuchika Ohta
January 2020

Contents

Chapter 1

Crystals: X-ray Crystal Structure Analysis

1.1 Discovery of X-ray

When you take an annual comprehensive medical examination like in Japanese schools and companies, radiography (X-ray examination) is used. X-ray was first discovered by German physicist Roentgen in 1895, therefore radiography is often called as Roentgen in Japan. This X-ray is an electromagnetic wave having a wavelength of about $1\,\text{Å}(0.1\,\text{nm} = 100\,\text{pm})$. X-ray was accidentally found by Roentgen, when he noticed that an invisible ray had the ability to expose a photographic dry plate. Since it was an unknown ray, it was named as X-ray from the association of an unknown quantity X in mathematics. Roentgen received the first Nobel Prize in Physics in 1901 for his discovery of X-ray.

1.2 X-ray Generation

Roentgen employed a bulb tube as illustrated in Fig. 1.1 in order to carry out a discharge experiment in vacuum. During this experiment, he noticed that a photographic dry plate was accidentally exposed, and that an unknown ray was emitted from this bulb tube. In this tube, he used a metal block called a target as the anode, and applied a high voltage to irradiate electrons from the cathode into the target. At that time, he found that this target emitted an unknown invisible light of X-ray.

As can be seen from Fig. 1.2, the intensity of the X-ray increases depending on the voltage applied between the anode and the

1

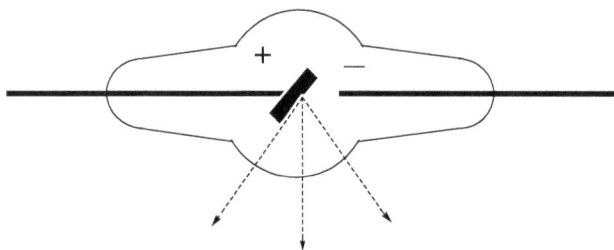

Figure 1.1.   Generation of X-ray. When a metal piece is placed as the anode in a high vacuum glass tube and irradiated by an electron beam at a high voltage from the cathode, X-ray is generated.

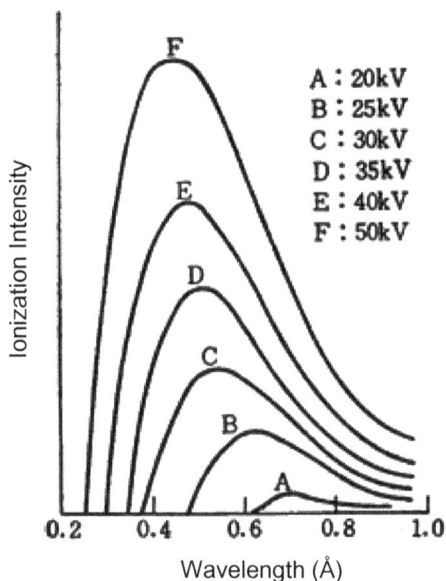

Figure 1.2.   Dependence of ionizition intensity and wavelength on the applied voltage in generation of X-ray.

cathode and the wavelength at the maximum intensity moves to the short wavelength. Moreover, the wavelength of X-rays is wide and continuous. Therefore, such X-rays are called "continuous X-rays" or "white X-rays." Although these X-rays are not colored, the term "white X-rays" is derived from the association that the visible lights become white when all seven color visible lights in the rainbow are combined.

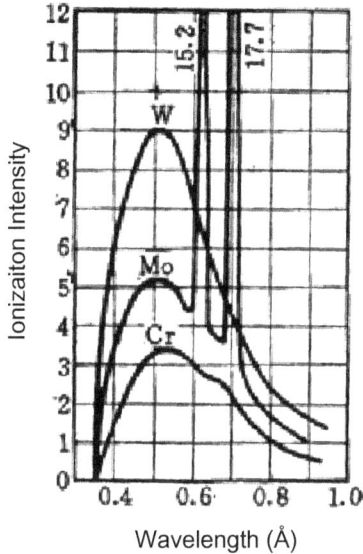

Figure 1.3. Emission of X-ray characteristic to the kinds of metal.

Very interestingly, depending on the type of target metal, a sharp X-ray with a wavelength specific to the metal appears like a bamboo shoot on a mountain of the white X-ray, as can be seen in Fig. 1.3. This metal-specific X-ray is called a characteristic X-ray. A metal often used for X-ray crystal structure analysis is copper, and the wavelength of its characteristic X-ray $K\alpha 1$ is $1.542\,\text{Å}$.

1.3 X-ray Diffraction

Since X-ray is one of the electromagnetic waves, it naturally has the nature of waves.

Let us consider here the nature of waves. Waves generally exhibit the properties of reflection, diffraction, and transmission. What appears significantly depends on the relative length between the wavelength, λ, and the interobject distance, d.

As shown in Fig. 1.4, when λ is much larger than d, i.e., $d \ll \lambda$, the reflection appears remarkably well. For example, it can be easily understood if you can imagine the wire mesh in the window glass of a microwave oven. The household microwave oven uses 2.45 GHz of

$$d$$

$$
\begin{array}{ll}
d \ll \lambda & \text{Reflection} \\
d \cong \lambda & \text{Diffraction} \\
d \gg \lambda & \text{Transmission}
\end{array}
$$

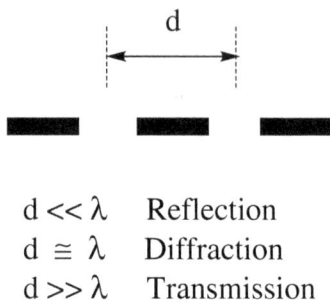

Figure 1.4. Wave phenomena dependent on relationship between wavelength and object spacings. When $d \ll \lambda$, **reflection** mainly occurs; e.g., metal-wire-mesh in a microwave oven window. When $d \approx \lambda$, **diffraction** mainly occurs; e.g., crystal lattice spacing $d \approx$ X-ray wavelength λ. When $d \gg \lambda$, **transmission** mainly occurs; e.g., radiograph.

microwaves that has $12.2\,\text{cm}$ of wavelength. Inside the window glass of the microwave oven, a wire mesh sheet is sealed, and the distance between the wire mesh is about $0.2\,\text{cm}$. The microwaves in the oven are reflected by the wire mesh and do not come out of the wire mesh. Therefore, even when you peep at the objects in the microwave oven through the wire mesh, your eyes will not be heated and cooked to become like boiled eggs. This is called **electrostatic shielding**.

On the other hand, when λ is much smaller than d, i.e., $d \gg \lambda$, the transmission also appears remarkably well. For example, in a microwave oven there is an indoor light and thanks to the light you can see what is in use. Since the indoor light used is visible light, the wavelength is $400\,\text{nm} \sim 800\,\text{nm}$ $(4 \times 10^{-6}\,\text{cm} \sim 8 \times 10^{-6}\,\text{cm})$. The wavelength is much shorter than $0.2\,\text{cm}$ of the wire mesh spacing. For this wavelength of electromagnetic waves, transmission appears remarkably well. Therefore, we can safely observe the objects in the microwave oven through the wire mesh.

When the λ is approximately the same as d, i.e., $d \approx \lambda$, the diffraction phenomenon also occurs remarkably well. For example, when in front of a light source such as a fluorescent light, the thumb and the forefinger are brought very close to each other and the light source is seen from the gap, bright and dark streaks appear alternately outside the gap, that is, inside the skin of both the fingers. In another case, when in the front of light source, two parallel pencils are brought very close to each other and you look at the light source through the narrow gap, bright and dark streaks appear alternately outside the gap, that is, inside the pencil. This striped pattern is a diffraction phenomenon of visible light that can be observed close to you. This is because the wavelength of visible light is 400 nm–800 nm (0.4 μm–0.8 μm), and the gap between two pencils is 0.01 mm–0.001 mm (10 μm–1 μm); the gap is almost the same as the wavelength of visible light, therefore such diffraction phenomena occurs — remember Young's Double Slits Experiment learned in high school physics.

You may see another example of diffraction phenomena in much longer wavelength in seashore: the sand on the shore is shaved down to the inside of the breakwater when the breakwaters on the shore are spaced every 10 m. This is also because a large diffraction phenomenon occurs when the wave wavelength and the breakwater spacing are nearly the same. Since the diffraction phenomenon is well-explained intuitively from the principle of Huygens in high school physics, the details are required to review. Thus, the diffraction phenomenon is universally seen in waves for all the lengths when d is nearly the same as λ.

The spacing of the arrangement of atoms, molecules, ions, etc., constituting the crystal is approximately in the order of Å (0.1 nm). Therefore, the relationship between d and λ of the X-ray corresponds just to the case of $d \approx \lambda$, thus when the X-ray is irradiated to the crystal, the diffraction phenomenon occurs remarkably.

As apparent from the above description, when X-ray is employed in a medical checkup, transmission phenomenon of the wave is used. On the other hand, when X-ray is employed for crystal structure analysis, diffraction phenomenon of the wave is used.

1.4 Bragg's Condition [1, 2]

Figure 1.5 schematically shows a crystal in which atoms or molecules are regularly and periodically arranged. When a crystal having such a periodic structure is irradiated with X-rays, diffraction occurs.

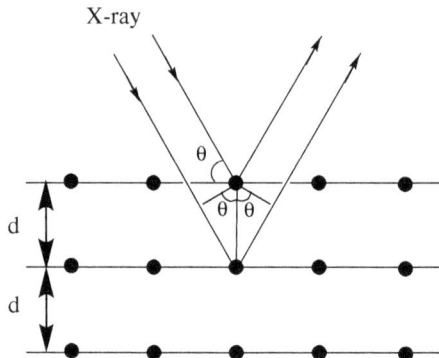

$$2d\sin\theta = n\lambda \quad n=1, 2, 3.......$$

Figure 1.5.   Bragg's law.

Hereupon, it is assumed that X-ray beams are reflected at atoms (or molecules) of the first and second layers in the crystal at an incident angle θ, as shown in this figure. Compared to the reflection from the first layer, the reflection from the second layer has an additional path difference of $2d\sin\theta$. When the path difference contains an integer number (n) of the waves, the reflection line appears as a thick line on the photographic plate because these waves have the same phase. However, when the path difference has a half integer $(n+1/2)$ of waves, the reflection line disappears on the photographic plate because these waves have the just opposite phase to cancel these intensities. Therefore, the condition for the appearance of the thick line is

$$2d\sin\theta = n\lambda \quad n = 1, 2, 3, \ldots$$

$$(1.1)$$

(d: spacing; θ: incidence angle; λ: wavelength of X-ray)

This is called as "Bragg's condition."

Such diffraction phenomena in X-rays were discovered by the father, Bragg, and his son, who won the Nobel Prize in Physics in 1915.

1.5 Crystal Lattice

Atoms, molecules, and/or ions in a crystal are three-dimensionally and regularly arranged in a three-dimensional space. This space arrangement is called a **space lattice**. From their symmetries, they are classified into 230 types of space groups.

As shown in Fig. 1.6, a parallelpipe composing with three vectors $\vec{a}, \vec{b}$ and $\vec{c}$ is called **unit lattice** or **unit cell**. The parameters a, b, c, α, β and γ are called **lattice constants**.

1.6 Seven Crystal Systems

The crystals have seven crystal systems, namely, cubic, tetragonal, orthorhombic rhombohedral (trigonal), hexagonal, monoclinic and triclinic. The lattice constants in each of the crystal systems are defined as shown in Table 1.1. You need to remember at least these seven crystal systems essential for research.

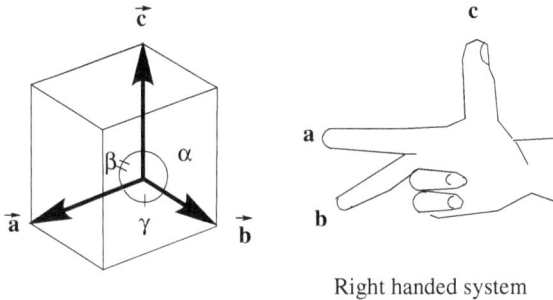

Right handed system

Figure 1.6.   Definition of crystal lattice constants of a, b, c, α, β and γ.

Table 1.1.   Seven crystal systems.

Crystal system	Crystal parameter
Cubic	$a = b = c, \alpha = \beta = \gamma = 90°$
Tetragonal	$a = b \neq c, \alpha = \beta = \gamma = 90°$
Orthorhombic	$a \neq b \neq c, \alpha = \beta = \gamma = 90°$
Rhombohedral	$a = b = c, \alpha = \beta = \gamma < 120°, \neq 90°$
Hexagonal	$a = b \neq c, \alpha = \beta = 90°, \gamma = 120°$
Monoclinic	$a \neq b \neq c, \alpha = \gamma = 90° \neq \beta$
Triclinic	$a \neq b \neq c, \alpha \neq \beta \neq \gamma$

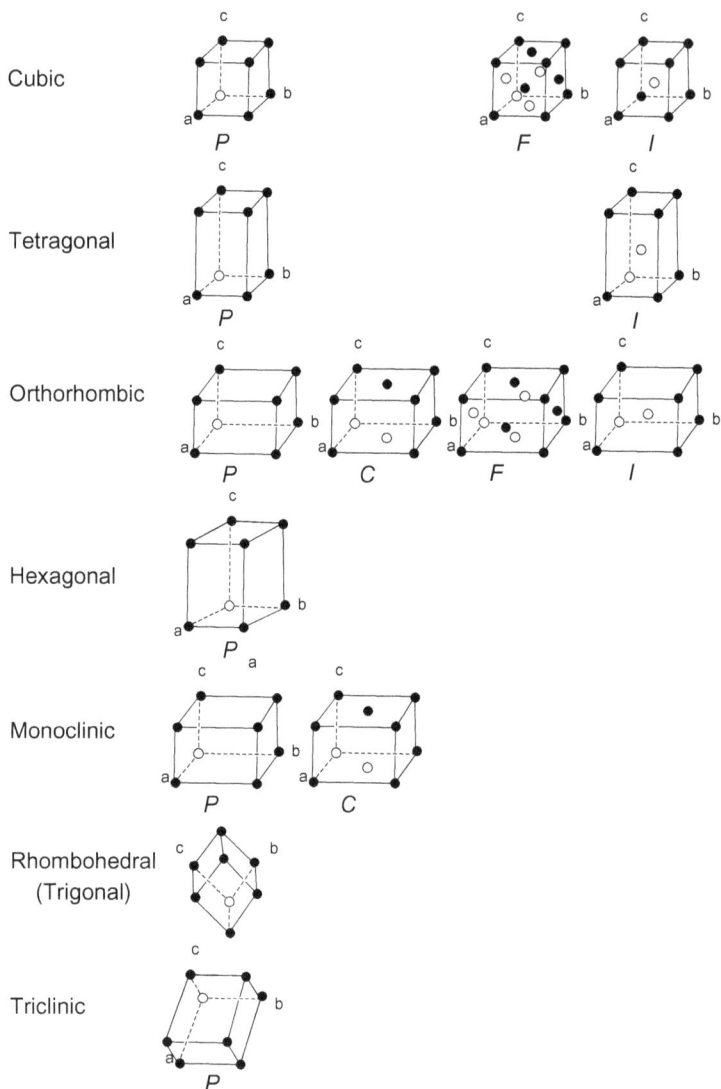

Figure 1.7.   Fourteen Bravais lattices.

It has also been mathematically proven that only **14 Bravais lattices** fill the space. For example, as can be seen from Fig. 1.7, the cubic system has three Bravais lattices: a simple cubic lattice (P), a face-centered cubic lattice (F), and a body-centered cubic lattice (I).

Therefore, crystal structure analysis using X-ray diffraction is, in short, to determine the lattice constants of the unit cell and identify its symmetry out of 230 space groups.

1.7 Miller Index

In Fig. 1.5, the X-ray reflection plane is taken horizontally for simplicity, but there should be various other reflection planes. As can be seen from Fig. 1.8, there are so many reflection planes that it is necessary to distinguish from which plane the reflection occurs. Therefore, "when a plane intersects at $\left(\frac{a}{h}, \frac{b}{k}, \frac{c}{l}\right)$ in the a, b and c axis, respectively, this plane including all the parallel planes is defined as a plane $(h\ k\ l)$. hkl is called the Miller index." However, when you only hear this definition, you may not understand it immediately. Therefore I will explain this Miller index by using simple examples of two-dimensional (2D) lattice and three-dimensional (3D) lattice, respectively, as follows. For easy comprehension, I will explain Miller index together with Weiss index. Weiss index is not currently used, but helps to understand the Miller index.

First of all, let us consider the case of a 2D lattice. In the case of Fig. 1.9[A], the straight line passes at 1 in the a-axis direction and 1 in the b-axis direction from the origin O. That is, the coefficient of (ab) is (11). This coefficient (11) is a plane index represented by the

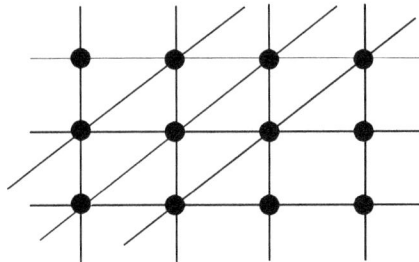

Figure 1.8. Miller Index. Since there are various reflective surfaces, they should be distinguished in some method. A plane intersecting at a/h, b/k, c/l and all the planes parallel to this one are named planes (hkl). The hkl is called as Miller index. (Weiss index is not currently used, but helps to understand Miller index.)

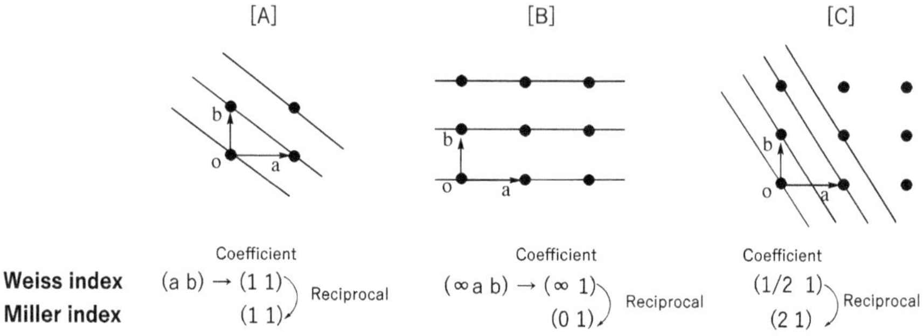

	[A]	[B]	[C]
	Coefficient	Coefficient	Coefficient
Weiss index	$(a\ b) \rightarrow (1\ 1)$	$(\infty\ a\ b) \rightarrow (\infty\ 1)$	$(1/2\ 1)$
Miller index	$(1\ 1)$	$(0\ 1)$	$(2\ 1)$

(with "Reciprocal" connecting each Weiss index to its Miller index)

Figure 1.9.   Weiss index and Miller index in the cases of some 2D lattices.

Weiss index. When the Weiss index (11) is changed to be reciprocal and then converted to an integer, the corresponding Miller index is obtained. Since this is $\left(\frac{1}{1}\ \frac{1}{1}\right)$, it becomes (11). This is not different from the Weiss index.

However, in the case of Fig. 1.9[B], this straight line passes at ∞ in the a-axis direction (= parallel to the a-axis) and at 1 in the b-axis direction from the origin O. That is, the coefficient of (ab) is $(\infty 1)$. This coefficient $(\infty 1)$ is a plane index represented by the Weiss index. When the Weiss index $(\infty 1)$ is changed to be reciprocal and then converted to an integer, the corresponding Miller index is obtained. Since this is $\left(\frac{1}{\infty}\ \frac{1}{1}\right)$, it becomes (01).

In the case of Fig. 1.9[C], this straight line passes at $\frac{1}{2}$ in the a-axis direction and at 1 in the b-axis direction from the origin O. Therefore, the Weiss index is $\left(\frac{1}{2}1\right)$. When the Weiss index $\left(\frac{1}{2}1\right)$ is changed to be reciprocal and then converted to an integer, the corresponding Miller index (21) is obtained.

As can be seen from these examples in Figs. 1.9[A–C], the Weiss index is intuitively easy to understand, but it is inconvenient because of the presence of infinity and fractions. Therefore, at present, the Miller index is used for the plane index, and the Weiss index is not used. However, the Weiss index is very useful for understanding the Miller index.

Next let us consider the case of a 3D lattice. In the case of Fig. 1.10[A], this plane passes at 1, 1 and 1 in the a, b and c axis

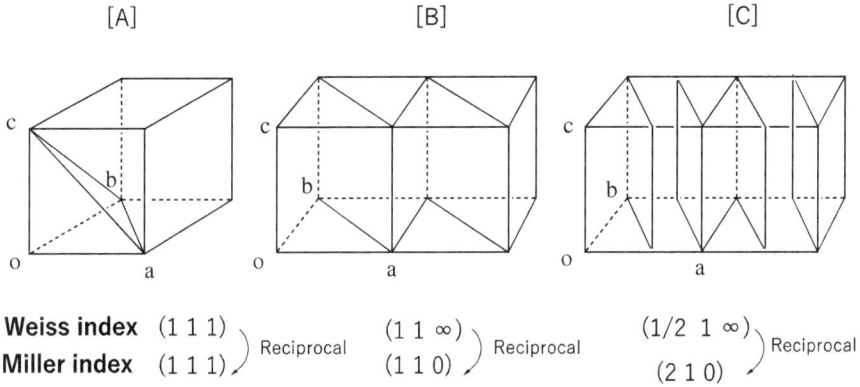

	[A]	[B]	[C]
Weiss index	(1 1 1)	(1 1 ∞)	(1/2 1 ∞)
Miller index	(1 1 1)	(1 1 0)	(2 1 0)

Reciprocal (for each column pair)

Figure 1.10. Weiss index and Miller index in the cases of some 3D lattices.

directions from the origin O, respectively. The coefficient of (abc) becomes (111). This coefficient (111) is a plane index represented by the Weiss index. When the Weiss index (111) is changed to be reciprocal and then converted to an integer, the corresponding Miller index (111) is obtained. This is not different from the Weiss index.

In the case of Fig. 1.10[B], this plane passes at 1, 1 and ∞ in the a, b and c axis directions (= parallel to the c-axis) from the origin O, respectively. The coefficient of (abc) is (11∞). This coefficient (11∞) is the Weiss index. When the Weiss index (11∞) is changed to be reciprocal and then converted to an integer, the corresponding Miller index (110) is obtained.

In the case of Fig. 1.10[C], this plane passes at $\left(\frac{1}{2} \ 1 \ \infty\right)$ in the a, b and c-axis directions from the origin O, respectively. Hence, when this Weiss index is $\left(\frac{1}{2} \ 1 \ \infty\right)$ is changed to be reciprocal and then converted to an integer, the corresponding Miller index (210) can be obtained. As can be seen from this figure in Fig. 1.10[C], all (210) planes are parallel to each other.

From the above-mentioned examples, especially from Figs. 1.9[C] and 1.10[C], you can finally understand the definition of Miller index: "when a plane intersects at $\left(\frac{a}{h}, \frac{b}{k}, \frac{c}{l}\right)$ in the a, b and c-axis respectively, this plane including all the parallel planes is defined as a plane ($h\ k\ l$). The hkl is called Miller index."

1.8 Relationship Among Spacing d, Miller Index and Lattice Constants

We consider the relationship among the interplanar spacing d, the Miller index, and the lattice constants.

1.8.1 *Simple 2D square lattice*

First, we will consider a spacing d in a simple 2D square illustrated in Fig. 1.11.

From Pythagorean theorem,

$$(2d)^2 = a^2 + a^2$$

$$\therefore \ \frac{1}{d^2} = \frac{2}{a^2} \tag{1.2}$$

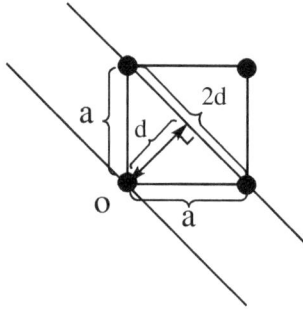

Figure 1.11. Relationship between a spacing d and a lattice constant a in a case of 2D square lattice.

Next, we consider the spacing d_{hk0} of a 2D square illustrated in Fig. 1.12. The d_{hk0} is the distance between ($hk0$) planes. Hereupon, from the similarity of the triangle,

$$\frac{a}{h} : x = \frac{b}{k} : d_{hk0} \tag{1.3}$$

$$\frac{a}{k} : y = \frac{a}{h} : d_{hk0} \tag{1.4}$$

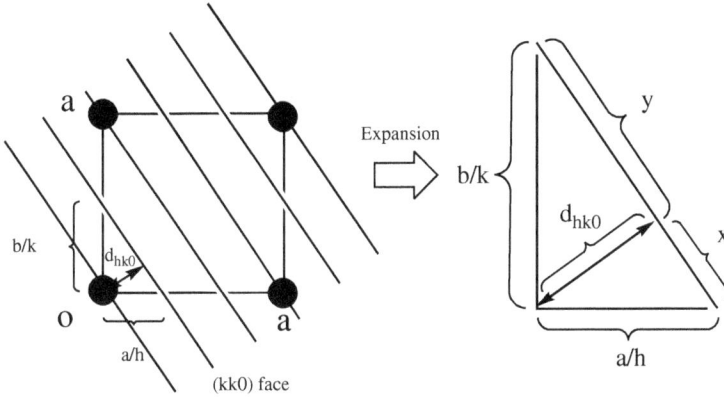

Figure 1.12. Relationship among a spacing d, Miller indices ($hk0$) and a lattice constant a in a case of 2D tetragonal (square) lattice.

From the Pythagorean theorem,

$$(x + y)^2 = \left(\frac{a}{h}\right)^2 + \left(\frac{a}{h}\right)^2 \tag{1.5}$$

When the parameters x and y are eliminated from these formulas (1.3)–(1.5), we obtain a formula,

$$\therefore \quad \frac{1}{(d_{hk0})^2} = \frac{h^2 + k^2}{a^2} \tag{1.6}$$

If $h = 1$ and $k = 1$ are entered into Equation (1.6), we can reproduce Equation (1.2). Accordingly, Equation (1.6) is more general than Equation (1.2), and Equation (1.2) holds at that special time.

1.8.2 *In the more general case of a 2D rectangle lattice*

The problem of calculating the spacing d_{hk0} in a 2D rectangle illustrated in Fig. 1.13 corresponds to calculation of the distance from one point to one straight line in a mathematical 2D vector space.

Therefore, we use the following two formulas described in the mathematical textbooks, for example, "Basic Mathematics for Engineers" by Ishihara Yano, pp. 146–147 [3].

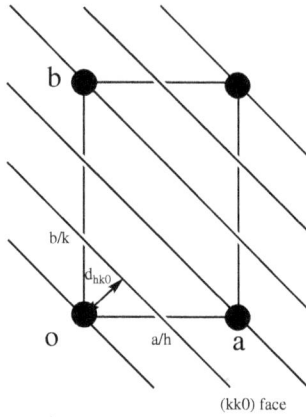

Figure 1.13.   Relationship among a spacing d, Miller indeces $(hk0)$ and lattice constants $(a$ and $b)$ in the case of a 2D rectangular lattice.

"Distance from one point to one straight line in a 2D plane"

- The equation of a straight line passing through two points $(a\ 0)$ and $(b\ 0)$ is

$$\frac{x}{a} + \frac{y}{b} = 1 \tag{1.7}$$

- The distance d from the point $P_0(x_0\ y_0)$ to the straight line $ax + by + c = 0$ is

$$d = \frac{|ax_0 + by_0 + c|}{\sqrt{a^2 + b^2}} \tag{1.8}$$

When $(x_0\ y_0) = (0\,0)$,

$$d = \frac{|c|}{\sqrt{a^2 + b^2}}$$

$$\therefore \ \frac{1}{d^2} = \frac{a^2 + b^2}{c^2} \tag{1.8'}$$

In Fig. 1.13, the line passes at two points, $\left(\frac{a}{h}\ 0\right)$ and $\left(\frac{b}{k}\ 0\right)$.

$$\frac{x}{\left(\frac{a}{h}\right)} + \frac{y}{\left(\frac{b}{k}\right)} = 1 \qquad (\because 1.7)$$

Therefore, when this equation is transformed,

$$\left(\frac{b}{k}\right)x + \left(\frac{a}{h}\right)y - \frac{ab}{hk} = 0$$

The distance from the origin to this straight line can be derived from Equation (1.8′)

$$\frac{1}{(d_{hk0})^2} = \frac{\left(\frac{b}{k}\right)^2 + \left(\frac{a}{h}\right)^2}{\left(\frac{ab}{hk}\right)^2}$$

To organize this,

$$\therefore \frac{1}{(d_{hk0})^2} = \frac{h^2}{a^2} + \frac{k^2}{b^2} \qquad (1.9)$$

(cf. S_E, Col_{rd} liquid crystal phase)

1.8.3 *In the case of a more general 3D orthorhombic lattice*

We will consider the distance from one point to one plane in a more general 3D orthorhombic lattice illustrated in Fig. 1.14. This also

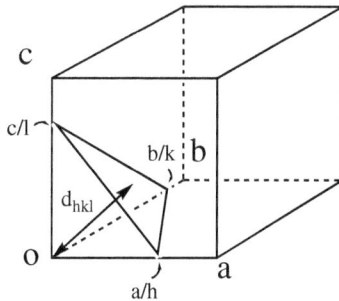

Figure 1.14. Relationship among a spacing d, Miller indices (hkl) and lattice constants $(a, b \text{ and } c)$ in the case of a 3D orthorhombic lattice.

corresponds to the calculation of the distance from one point to one plane in a mathematical 3D vector space.

Therefore, we again use the following two formulas described in the mathematical textbook, "Basic Mathematics for Engineers" by Ishihara Yano pp. 152–153 [3].

"Distance from one point to one plane in a 3D space"

- The equation of the plane passing through three points $(a\ 0\ 0)$, $(0\ b\ 0)$, $(0\ 0\ c)$ is

$$\frac{x}{a} + \frac{y}{b} + \frac{z}{c} = 1 \tag{1.10}$$

which is transformed into

$$bcx + acy + abz - abc = 0 \tag{1.10'}$$

- The distance d from the point $P_0(x_0\ y_0\ z_0)$ to the plane $ax + by + cz + e = 0$ is

$$d = \frac{|ax_0 + by_0 + cz_0 + e|}{\sqrt{a^2 + b^2 + c^2}} \tag{1.11}$$

When $(x_0\ y_0\ z_0) = (0\ 0\ 0)$,

$$d = \frac{|e|}{\sqrt{a^2 + b^2 + c^2}}$$

$$\therefore \frac{1}{d^2} = \frac{a^2 + b^2 + c^2}{e^2} \tag{1.11'}$$

When the plane is described as ⑩′, ⑪′ can be derived as

$$\frac{1}{d^2} = \frac{(bc)^2 + (ac)^2 + (ab)^2}{(abc)^2}$$

$$\therefore \frac{1}{d^2} = \frac{1}{a^2} + \frac{1}{b^2} + \frac{1}{c^2} \tag{1.12}$$

In Fig. 1.14, the plane passes at three points, $(\frac{a}{h}00)$, $(0\frac{b}{k}0)$, $(00\frac{c}{l})$.

$$\frac{1}{(d_{hkl})^2} = \frac{1}{(\frac{a}{h})^2} + \frac{1}{(\frac{b}{k})^2} + \frac{1}{(\frac{c}{l})^2} \qquad \therefore (1.12)$$

$$\therefore \frac{1}{(d_{hkl})^2} = \frac{h^2}{a^2} + \frac{k^2}{b^2} + \frac{l^2}{c^2} \qquad (1.13)$$

Relational equation for orthorhombic crystal [4] cf. Table 1.2.

In the same way, the distance from one point to one plane in cubic and tetragonal crystal systems can be derived. In the cases of other crystal systems whose angle is not 90°, it is much more difficult but

Table 1.2. Relationship among spacing d_{hkl}, Miller indices and lattice constants in 3D-crystal systems.

Crystal system	Spacing d_{hkl}
Cubic	$\dfrac{1}{d_{hkl}^2} = \dfrac{h^2 + k^2 + l^2}{a^2}$
Tetragonal	$\dfrac{1}{d_{hkl}^2} = \dfrac{h^2 + k^2}{a^2} + \dfrac{l^2}{c^2}$
Orthorhombic	$\dfrac{1}{d_{hkl}^2} = \dfrac{h^2}{a^2} + \dfrac{k^2}{b^2} + \dfrac{l^2}{c^2}$
Hexagonal	$\dfrac{1}{d_{hkl}^2} = \dfrac{4}{3}\left(\dfrac{h^2 + hk + k^2}{a^2}\right) + \dfrac{l^2}{c^2}$
Monoclinic	$\dfrac{1}{d_{hkl}^2} = \dfrac{1}{\sin^2 \beta}\left(\dfrac{h^2}{a^2} + \dfrac{k^2 \sin^2 \beta}{b^2} + \dfrac{l^2}{c^2} - \dfrac{2hl\cos \beta}{ac}\right)$
Rhombohedral	$\dfrac{1}{d_{hkl}^2} = \dfrac{(h^2 + k^2 + l^2)\sin^2 \alpha + 2(hk + kl + hl)(\cos^2 \alpha - \cos \alpha)}{a^2(1 - 3\cos^2\alpha + 2\cos^2 \alpha)}$
Triclinic	$\dfrac{1}{d_{hkl}^2} = \dfrac{1}{V^2}(S_{11}h^2 + S_{22}k^2 + S_{33}l^2 + 2S_{12}hk$ $+ 2S_{23}kl + 2S_{13}hl)$ $V^2 = a^2b^2c^2(1 - \cos^2 \alpha - \cos 2\beta - \cos 2\gamma + 2\cos \alpha \cos \beta \cos \gamma)$ $S_{11} = b^2c^2 \sin^2 \alpha, \; S_{22} = a^2c^2 \sin^2 \beta, \; S_{33} = a^2b^2 \sin^2 \gamma,$ $S_{12} = abc^2(\cos \alpha \cos \beta)$

can be derived as the mathematical vector problems. Accordingly, "the relationship among the interplanar spacing d, the Miller index and the lattice constants" in seven crystal systems can be obtained as shown in Table 1.2.

In the next step, we will look at the actual method how to carry out X-ray crystal structure analysis by using the relationships shown in Table 1.2. This book will give an overview. For further details, you are required to refer to a specialized book, for example, Toshio Sakurai's, *Guide to X-ray Crystallographic Analysis* [2]. In X-ray crystal structure analysis, there are single crystal method and powder crystal method.

1.9   Single Crystal Method

For the single crystal method, a Weisenberg camera, a precession camera, or a four-axis diffractometer is used. The current mainstream is the four-axis diffractometer, but until a short time ago, Weisenberg camera and precession camera have been actively used. Therefore, when you will read the past papers, you will need the knowledge of these camera methods. In the following sections, an overview is presented on all these old and new methods.

1.9.1   *Weisenberg camera*

Figure 1.15 shows a schematic view of the Weisenberg camera device. A single crystal is set at the tip of a needle, X-ray is applied to it, and its diffraction image is recorded with a cylindrical film around it. However, in order to irradiate X-rays to all parts of the single crystal, the single crystal is rotated and, at the same time, the cylindrical film is vibrated left and right.

This very sophisticated mechanism employs the mechanism of the thread winding machine invented by Leonardo da Vinci. The purpose of vibrating the rod of the thread winding machine to the left and right is to wind a yarn of uniform thickness around the rod [5]. Thus, the superior spinning mechanism of the textile machine is applied to this Weisenberg camera device. It is very interesting that Weisenberg had so extensive knowledge of textile machinery,

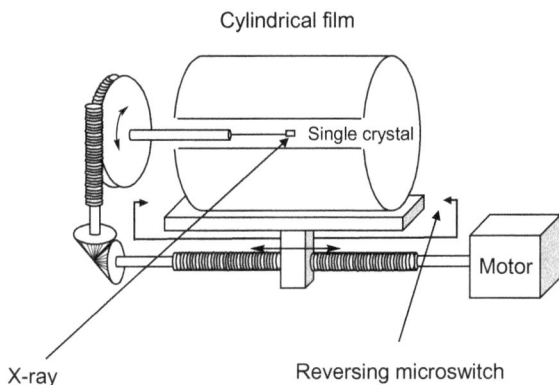

Figure 1.15.   Weisenberg camera.

apparently unrelated to X-ray single crystal structure analysis, that led to the development of this fantastic device.

The actual Weisenberg photograph gives severely distorted spot sequences, as shown in Fig. 1.16[A]. A translucent analysis sheet such as Fig. 1.16[B] is placed on this photograph, and the points on the intersections are read to index. Therefore, it is extremely complicated and time-consuming.

1.9.2   *Precession camera*

Since the severely distorted spot sequences are difficult to analyze, the precession camera is designed to align spots in straight lines, as can be seen from an example of the precession photograph shown in Fig. 1.17. Here we can see the undistorted hexagonal spot sequences. Accordingly, it becomes much easier for the indexation and determination which spots are reflected from which (*hkl*) plane.

1.9.3   *Four-axis diffractometer*

Recently, the four-axis diffractometer has become dominant. Its schematic diagram is shown in Fig. 1.18. Three axes are required to rotate the crystal in all directions, and one axis is required to determine the position of the Geiger counter, so that a total of four axes of rotation control will be able to collect all the reflections

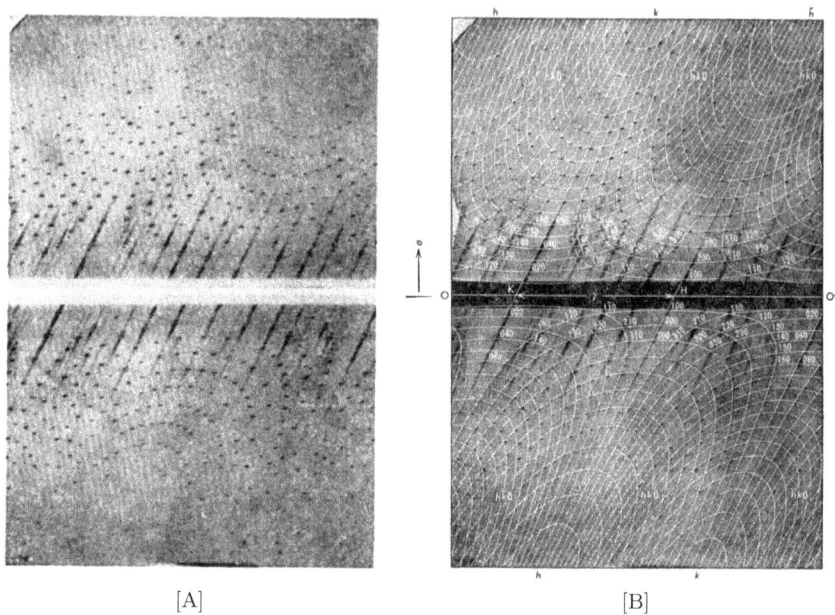

[A] [B]

Figure 1.16. An example of Weisenberg photograph [A] and the indexation [B]. Reproduced from H. Shimanouchi, Y. Sasada, *Bull. Chem. Soc. Jpn.*, **42**, 334 (1969), with kind permission of the Chemical Society of Japan.

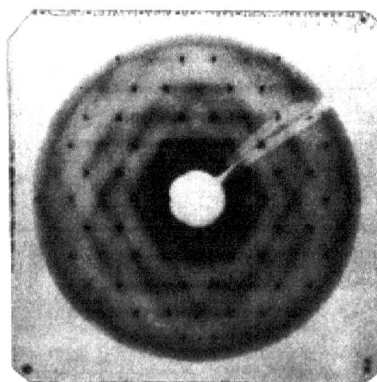

Figure 1.17. Precession photograph of AgI at rt. Reproduced from K. Koto, Nihon Kessyo Gakkaishi, 38, 138–143 (1996), with kind permission of the Nihon Kessyo Gakkai.

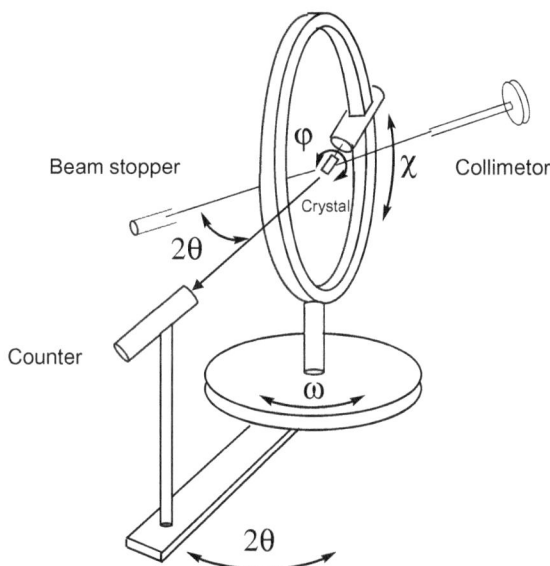

Figure 1.18. Schematic illustration of the four-circle X-ray diffractometer. Four circles: ϕ, χ, θ and ω.

from the entire space around the crystal. Therefore, this X-ray apparatus is called a four-axis diffractometer. With the extremely rapid development of computers, the position and intensity of spots (reflections) can be more easily and automatically measured by this apparatus, and the analysis can be more extremely labor-saving. Thus, X-ray single crystal structure analysis has become much more easier to carry out for many people who are not experts.

1.9.4   *What can be revealed by X-ray single crystal structure analysis*

As shown in Fig. 1.19, molecular structure and crystal structure can be revealed from X-ray single crystal structure analysis. Diagrams [A], [B] and [C] show the periodic crystal lattice, the molecular structure, and the crystal structure formed by the molecules arranged in the lattice, respectively. Thus, both molecular and crystalline structures are revealed from X-ray single crystal structure analysis.

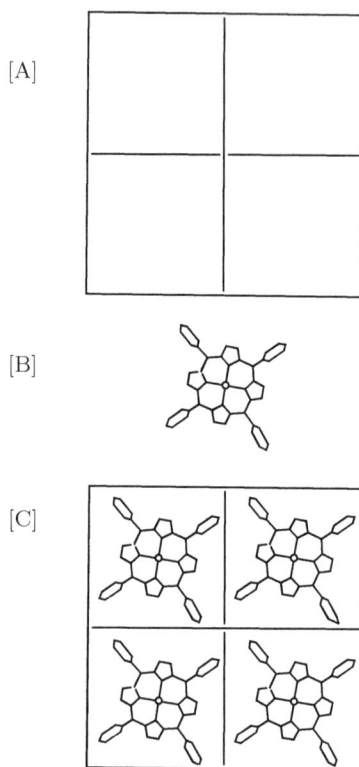

Figure 1.19. What one can reveal from X-ray single crystal structural analysis is both molecular structure and crystal structure: [A] periodic lattice, [B] molecular structure, and [C] crystal structure formed by arranging molecules in the lattice.

1.10 Powder Crystal Method

If a good single crystal is obtained, the single crystal structure analysis is possible as described above. However, if a good single crystal having an appropriate size cannot be obtained, and if only powder crystals can be obtained beyond any efforts, you only have to instead employ another powder crystal method shown in Fig. 1.20. In this figure, both [A] and [B] are the photo methods, and [B] is the counter method for the powder crystal method.

In the photo methods, powder crystals are filled in a narrow and thin glass tube (capillary) to serve a measurement sample as

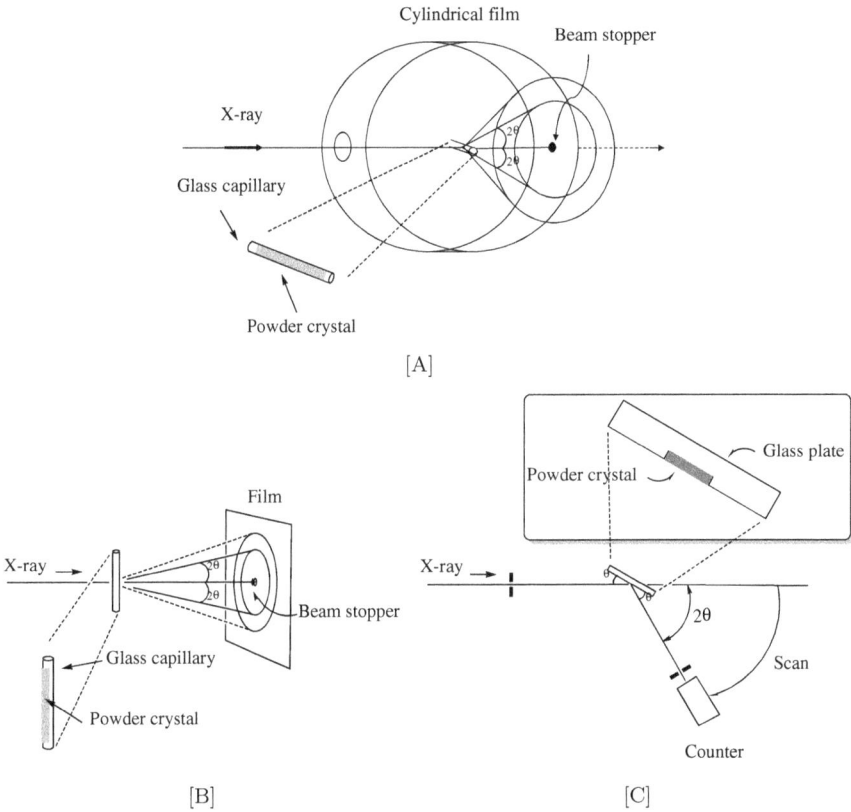

Figure 1.20. Powder crystal methods: [A] for a strip-shaped cylindrical film, [B] for a square plain film and [C] for a Geiger counter.

shown in the figure. The diffraction image is recorded as a strip-shaped cylindrical film in the figure [A] or a square plain film in the figure [B].

On the other hand, in the powder crystal method [C], powder crystals are placed on a hollow of a glass plate and rubbed off till reaching the level of glass surface by using another glass plate to obtain a measurement sample. For the X-ray diffraction, a Geiger counter is set at the 2θ position and scanned to measure the angles and intensities at which the diffractions occur, as illustrated in this figure. This counter method is convenient because it does not require development of the photograph.

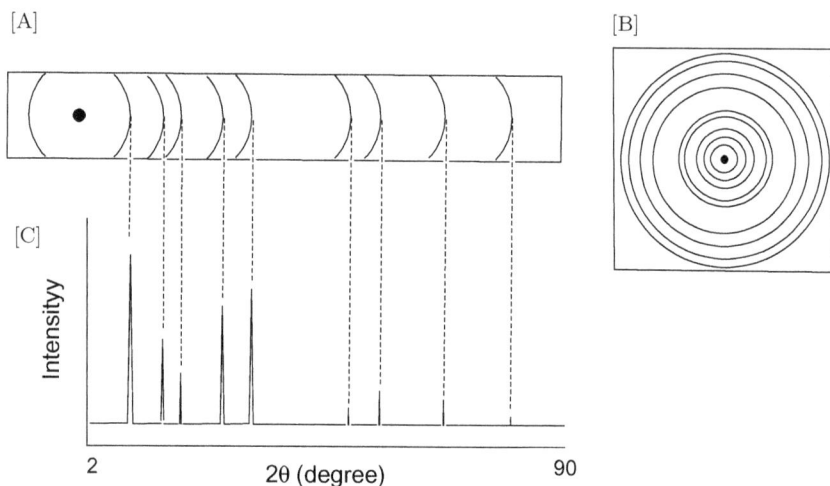

Figure 1.21.   X-ray diffraction profiles: [A] for a strip-shaped cylindrical film, [B] for a square plain film and [C] for a Geiger counter.

The powder photograph techniques [A] and [B] give the concentric diffraction images like [A] and [B] in Fig. 1.21. These diffraction patterns well correspond to the diffraction pattern in [C]. They can be converted to each other by calculation (cf. Exercise 1.9).

In the following sections, the reciprocal lattices and extinction rules will be explained. The biggest hurdle in learning X-ray crystallography is to understand the concept of reciprocal lattices and extinction rules. First of all, we will consider the reciprocal lattices. The following thinking method is the easiest way to understand the reciprocal lattices.

1.11   Reciprocal Lattice

What is a reciprocal lattice? Table 1.3 summarizes the relationship between the real lattice and the reciprocal lattice in three simple crystal systems: orthogonal, monoclinic, and hexagonal crystals. For these crystals, we think about what real lattices and reciprocal lattices are.

Roughly speaking, it is easy to understand that the real lattices can be considered as visible lattices. For example, if real lattices (unit

Table 1.3. Relationship between crystal lattices and reciprocal lattices.

In the simplest cases

(1) In the case that three angles are right (orthogonal system)

$$V = abc \qquad\qquad V^* = a^*b^*c^*$$

$$a = \frac{1}{a^*}, \, b = \frac{1}{b^*}, \, c = \frac{1}{c^*} \qquad\qquad a^* = \frac{1}{a}, \, b^* = \frac{1}{b}, \, c^* = \frac{1}{c}$$

$$\alpha = \beta = \gamma = 90° \qquad\qquad \alpha^* = \beta^* = \gamma^* = 90°$$

(2) In the case that two angles are right (monoclinic system)

$$V = abc \sin\beta \qquad\qquad V^* = a^*b^*c^* \sin\beta^*$$

$$a = \frac{1}{a^* \sin\beta^*}, \, b = \frac{1}{b^*}, \, c = \frac{1}{c^* \sin\beta^*} \qquad a^* = \frac{1}{a \sin\beta}, \, b^* = \frac{1}{b}, \, c^* = \frac{1}{c \sin\beta}$$

$$\alpha = \gamma = 90°, \, \beta = 180° - \beta^* \qquad\qquad \alpha^* = \gamma^* = 90°, \, \beta^* = 180° - \beta$$

(3) In the case that two angles are right and the one angle is 120°
(hexagonal system)

$$V = \frac{\sqrt{3}}{2}abc \qquad\qquad V^* = \frac{\sqrt{3}}{2}a^*b^*c^*$$

$$a = b = \frac{2}{a^*\sqrt{3}} = \frac{2}{b^*\sqrt{3}}, \, c = \frac{1}{c^*} \qquad a^* = b^* = \frac{2}{a\sqrt{3}} = \frac{2}{b\sqrt{3}}, \, c^* = \frac{1}{c}$$

$$\alpha = \beta = 90°, \, \gamma = 120° \qquad\qquad \alpha^* = \beta^* = 90°, \, \gamma^* = 180° - \gamma = 60°$$

cells) are stacked three-dimensionally to form a real crystal, you can see the shape like as quartz or calcite single crystal in your real life. Hence, you can easily imagine that the real lattices lead to the visible real crystals.

As can be seen from Table 1.3, the reciprocal lattice represents the space formed by using the reciprocals of the lattice constants and the complement angles. As shown in Fig. 1.22, when a small single crystal is placed at the center of a large sphere and this crystal is irradiated with X-ray beams, the X-ray beams are diffracted out of this crystal to draw an image on the inner wall of the sphere. The image is a reciprocal lattice. Briefly speaking, a reciprocal lattice is an imaginary lattice exposed on a photographic film attached to the inner wall of this sphere. It is a space drawn by the reciprocals such as $1/a$ and $1/b$ and the complements such as $\gamma^* = 180 - \gamma$. If the real lattice constant is long, the corresponding reciprocal lattice constant is short; if the real lattice is short, the corresponding reciprocal lattice

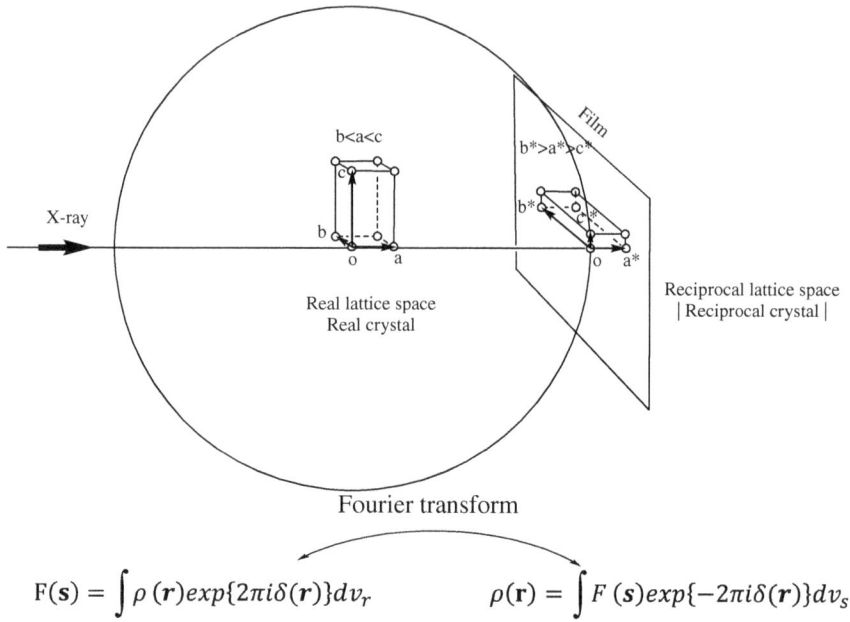

Figure 1.22.   Relationship between a real lattice and the reciprocal lattice in the case of an orthorhombic crystal.

constant is long. Thus, you can understand that the reciprocal lattice is considered as a virtual space drawn by the reciprocals of the lattice constants and the complement angles.

Therefore, X-ray structural analysis can be said as conversion from reciprocal lattice (virtual image) to real lattice (real image). For details of the relationship between 3D real lattices and reciprocal lattices, refer to a textbook on X-ray single crystal structure analysis, for example, "Guide to X-ray Crystal Analysis" written by Toshio Sakurai [2].

1.12   Structure Factor and Extinction Rules

1.12.1   *Structure factor*

When using a monochromatic (characteristic) X-ray (*e.g.*, Cu Kα: $\lambda = 1.542$ Å), the following two equations hold:

$$c = \nu\lambda \tag{1.14}$$

(Amplitude of X-ray scattering at frequency ν)

$$\propto \text{(electron density } (\rho) \text{ in crystal)} \qquad (1.15)$$

Wave equation of the scattered X-ray is presented as:

$$\rho(\mathbf{r}) \exp\{2\pi i(\nu t + \delta(\mathbf{r}))\} \qquad (1.16)$$

(expression for general waves) here,

$$2\pi\delta(\mathbf{r}) = \text{(phase)} \qquad (1.17)$$

Since superposition of the scattered X-rays ($\Rightarrow$ superposition of the diffracted X-rays) is an integral over the entire bond of Equation (1.16),

$$E = \int \rho(\mathbf{r}) \exp\{2\pi i(\nu t + \delta(\mathbf{r}))\} dv \qquad (1.18)$$

When this equation is divided into a part containing time and a part not containing time, it is divided into a part related only to the crystal structure and a part related only to the X-ray vibration.

$$\therefore \ E = \int \rho(\mathbf{r}) \exp\{2\pi i\delta(\mathbf{r})\} dv \cdot \exp(2\pi i\nu t)$$

$$= F \cdot \exp(2\pi i\nu t) \qquad (1.19)$$

$$F = \int \rho(\mathbf{r}) \exp\{2\pi i\delta(\mathbf{r})\} dv \qquad (1.20)$$

F is a part related only to the structure of the crystal, which is called a **structure factor**.

Since the wave intensity is defined as squared amplitude, the intensity of the diffraction line is

$$EE^* = F \cdot \exp(2\pi i\nu t) \cdot F^* \cdot \exp(-2\pi i\nu t) = FF^* = |F|^2 \ \ (1.21)$$

Next, the relationship among Equations (1.20), (1.21) and Miller index is derived.

For details, refer to a textbook for X-ray single crystal structure analysis, such as *Guide to X-ray Crystal Analysis* [2] by Toshio Sakurai.

The relationship between the structure factor of Equation (1.20) and Miller index is as follows.

Crystal structure factor:

$$F(hkl) = \sum_{i}^{N} f_i \cdot \exp\{2\pi i(hx_i + ky_i + lz_i)\} \tag{1.22}$$

Where N is the number of atoms (molecules) in the unit cell, f_i is the scattering power of the i-th atom (molecule), (hkl) is the Miller index, and (xyz) is Cartesian coordinates. Temperature factors are excluded.

Therefore, the diffraction intensity I(hkl) at this time is

$$I(hkl) = |F(hkl)|^2 \tag{1.23}$$

1.12.2   *Extinction rules*

Extinction rules mean that special combinations of Miller indices (hkl) result in extinction of the diffraction intensity I(hkl): the sum of the phases of the diffraction lines becomes zero depending on the symmetry of the crystal structure. As can be seen from Equation (1.23), when the diffraction intensity I(hkl) is zero, the crystal structure factor F(hkl) is also zero. Therefore, extinction rule can be derived from the crystal structure factor F(hkl).

Hereupon, we consider the extinction rule in the case of NaCl having a face-centered cubic lattice, as an example. The unit cell of NaCl is illustrated in Fig. 1.23. In this figure, black and white circles indicate Na$^+$ and Cl$^-$ ions, respectively. Table 1.4 shows the Cartesian coordinates (xyz) of all of the Na$^+$ and Cl$^-$ ions, the number n of the ions contained in the unit cell, and the value of $\exp\{2\pi i(hx_i + ky_i + lz_i)\}$. As can be seen from this table, the sum F(hkl) is:

$$F(hkl) = f_{Na}[1 + (-1)^{h+k} + (-1)^{l+h} + (-1)^{k+l}]$$
$$+ f_{Cl}[(-1)^h + (-1)^k + (-1)^l + (-1)^{h+k+l}] \tag{1.24}$$

From this equation, the following rules can be derived.

a (c)

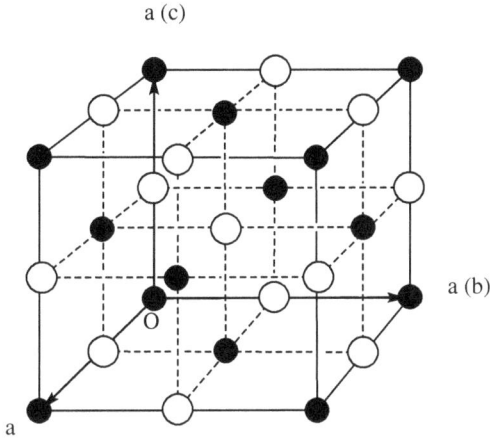

Figure 1.23. Unit cell of NaCl.

When h, k and l are all even, $F(hkl) = 4(f_{Na} + f_{Cl}) \rightarrow$ the intensity is strong.

When h, k and l are all odd, $F(hkl) = 4(f_{Na} - f_{Cl}) \rightarrow$ the intensity is weak.

When one is odd and the other are even, or when one is even and the others are odd $\rightarrow F(hkl) = 0$ ($=$ Extinction rule of face-centered cubic lattice.)

Thus, the extinction rule can be obtained from the crystal structure factor of Equation (1.22).

There are extinction rules corresponding to 230 space groups. (Reference: Toshio Sakurai, *Guide to X-ray Crystallography*. pp. 258–267 [2].) Thus, the space group can be determined from these extinction rules.

1.13 Simple Example of Crystal Structure Analysis

Figure 1.24 shows the X-ray diffraction pattern of NaCl measured by the counter method for the powder crystal (Fig. 1.20[C]). This is subjected to crystal structure analysis according to the following procedures.

Table 1.4.   Derivation of the crystal structure factor F(hkl) for a face-centered cubic lattice of NaCl.

Coordinate of Na	n in nf_{Na}	$\exp\{2\pi i(hx_i + ky_i + lz_i)\}$	Coordinate of Cl	n in nf_{Cl}	$\exp\{2\pi i(hx_i + ky_i + lz_i)\}$
$(0\,0\,0)$	$\frac{1}{8}$	$\exp\{0\} = 1$	$\left(\frac{1}{2}\,0\,0\right)$	$\frac{1}{4}$	$\exp\left\{2\pi i\left(\frac{1}{2}h\right)\right\} = (-1)^h$
$(1\,0\,0)$	$\frac{1}{8}$	$\exp\{2\pi i(h)\} = 1$	$\left(\frac{1}{2}\,\frac{1}{2}\,0\right)$	$\frac{1}{4}$	$\exp\left\{2\pi i\left(h + \frac{1}{2}k\right)\right\} = (-1)^k$
$(1\,1\,0)$	$\frac{1}{8}$	$\exp\{2\pi i(h+k)\} = 1$	$\left(\frac{1}{2}\,1\,0\right)$	$\frac{1}{4}$	$\exp\left\{2\pi i\left(\frac{1}{2}h + k\right)\right\} = (-1)^h$
$(0\,1\,0)$	$\frac{1}{8}$	$\exp\{2\pi i(k)\} = 1$	$\left(0\,\frac{1}{2}\,0\right)$	$\frac{1}{4}$	$\exp\left\{2\pi i\left(\frac{1}{2}k\right)\right\} = (-1)^k$
$\left(\frac{1}{2}\,\frac{1}{2}\,0\right)$	$\frac{1}{2}$	$\exp\left\{2\pi i\left(\frac{1}{2}h + \frac{1}{2}k\right)\right\} = (-1)^{h+k}$	$\left(0\,0\,\frac{1}{2}\right)$	$\frac{1}{4}$	$\exp\left\{2\pi i\left(\frac{1}{2}l\right)\right\} = (-1)^l$
$\left(\frac{1}{2}\,0\,\frac{1}{2}\right)$	$\frac{1}{2}$	$\exp\left\{2\pi i\left(\frac{1}{2}h + \frac{1}{2}l\right)\right\} = (-1)^{h+l}$	$\left(1\,0\,\frac{1}{2}\right)$	$\frac{1}{4}$	$\exp\left\{2\pi i\left(h + \frac{1}{2}l\right)\right\} = (-1)^l$
$\left(1\,\frac{1}{2}\,\frac{1}{2}\right)$	$\frac{1}{2}$	$\exp\left\{2\pi i\left(h + \frac{1}{2}k + \frac{1}{2}l\right)\right\} = (-1)^{k+l}$	$\left(1\,1\,\frac{1}{2}\right)$	$\frac{1}{4}$	$\exp\left\{2\pi i\left(h + \frac{1}{2}l\right)\right\} = (-1)^l$
$\left(\frac{1}{2}\,1\,\frac{1}{2}\right)$	$\frac{1}{2}$	$\exp\left\{2\pi i\left(\frac{1}{2}h + k + \frac{1}{2}l\right)\right\} = (-1)^{h+l}$	$\left(0\,1\,\frac{1}{2}\right)$	$\frac{1}{4}$	$\exp\left\{2\pi i\left(k + \frac{1}{2}l\right)\right\} = (-1)^l$
$\left(0\,\frac{1}{2}\,\frac{1}{2}\right)$	$\frac{1}{2}$	$\exp\left\{2\pi i\left(\frac{1}{2}k + \frac{1}{2}l\right)\right\} = (-1)^{k+l}$	$\left(\frac{1}{2}\,\frac{1}{2}\,\frac{1}{2}\right)$	1	$\exp\left\{2\pi i\left(\frac{1}{2}h + \frac{1}{2}k + \frac{1}{2}l\right)\right\} = (-1)^{h+k+l}$

(Continued)

Table 1.4. (*Continued*)

Coordinate of Na	n in nf_{Na}	$\exp\{2\pi i(hx_i + ky_i + lz_i)\}$	Coordinate of Cl	n in nf_{Cl}	$\exp\{2\pi i(hx_i + ky_i + lz_i)\}$
$(0\ 0\ 1)$	$\frac{1}{8}$	$\exp\{2\pi i\,(l)\} = 1$	$\left(\frac{1}{2}\ 0\ 1\right)$	$\frac{1}{4}$	$\exp\left\{2\pi i\left(\frac{1}{2}h + l\right)\right\} = (-1)^h$
$(1\ 0\ 1)$	$\frac{1}{8}$	$\exp\{2\pi i\,(h + l)\} = 1$	$\left(1\ \frac{1}{2}\ 1\right)$	$\frac{1}{4}$	$\exp\left\{2\pi i\left(h + \frac{1}{2}k + l\right)\right\} = (-1)^k$
$(1\ 1\ 1)$	$\frac{1}{8}$	$\exp\{2\pi i\,(h + k + l)\} = 1$	$\left(\frac{1}{2}\ 1\ 1\right)$	$\frac{1}{4}$	$\exp\left\{2\pi i\left(\frac{1}{2}h + k + l\right)\right\} = (-1)^h$
$(0\ 1\ 1)$	$\frac{1}{8}$	$\exp\{2\pi i\,(k + l)\} = 1$	$\left(0\ \frac{1}{2}\ 1\right)$	$\frac{1}{4}$	$\exp\left\{2\pi i\left(\frac{1}{2}k + l\right)\right\} = (-1)^k$
$\left(\frac{1}{2}\ \frac{1}{2}\ 1\right)$	$\frac{1}{2}$	$\exp\left\{2\pi i\left(\frac{1}{2}h + \frac{1}{2}k + l\right)\right\} = (-1)^{h+k}$	Sum $= f_{Cl}\left[\dfrac{1}{4} \times 4(-1)^h + \dfrac{1}{4} \times 4(-1)^k + \dfrac{1}{4} \times 4(-1)^{hl} + (-1)^{h+k+l}\right]$		

$$= f_{Cl}\left[(-1)^h + (-1)^k + (-1)^l + (-1)^{h+k+l}\right]$$

$$\text{Sum} = f_{Na}\left[\frac{1}{8} \times 8(1) + \frac{1}{2} \times 2(-1)^{h+k} + \frac{1}{2} \times 2(-1)^{h+l} \right.$$
$$\left. + \frac{1}{2} \times 2(-1)^{k+l}\right]$$

$$= f_{Na}[1 + (-1)^{h+k} + (-1)^{h+l} + (-1)^{k+l}]$$

Therefore, the sum total $F(hkl)$ is

$$F(hkl) = f_{Na}\left[1 + (-1)^{h+k} + (-1)^{h+l} + (-1)^{k+l}\right] + f_{Cl}\left[(-1)^h + (-1)^k + (-1)^l + (-1)^{h+k+l}\right]$$

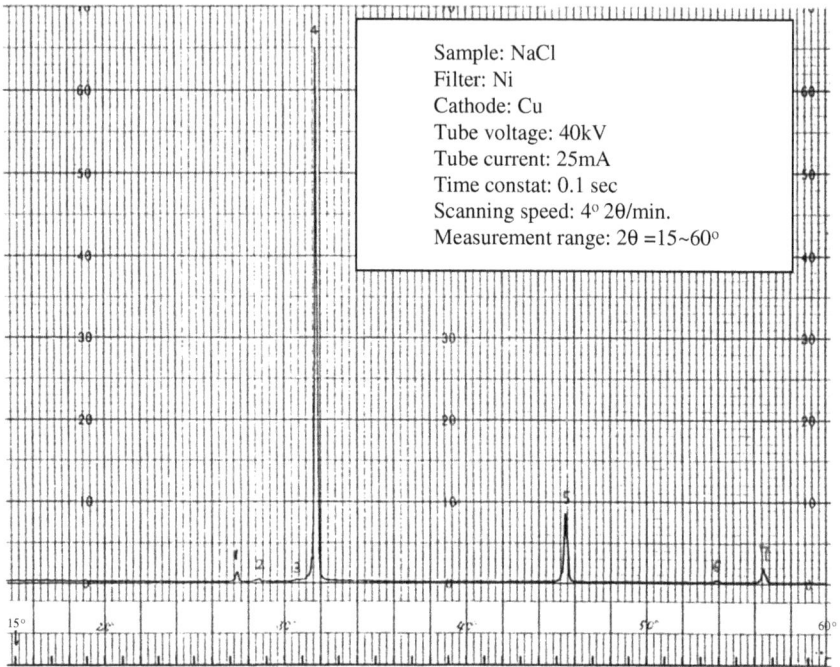

Figure 1.24.   Diffraction pattern of powder crystal of NaCl.

Procedure 1

First, draw a vertical line from the top of each of the seven peaks observed in the X-ray diffraction pattern in this figure, read the angle of the horizontal axis 2θ, and obtain the interplanar spacing d (Å) of each diffraction line by using the Bragg's condition (Equation (2.1): $\lambda = 1.5418$ Å of Cu $K\alpha$). Also, the relative intensity ratio (I/I_1) of those peaks is calculated. The relative intensity ratio represents the height of each peak as a percentage when the height of the strongest peak (I_1) as 100. These results are summarized in Table 1.5.

Procedure 2

It is already known from microscopic observation that crystal of NaCl has cubic system. With this in mind, each diffraction line is indexed.

Table 1.5.   The observed values of X-ray diffraction for NaCl.

Peak No.	2θ (degree)	Spacing $d(\text{Å})$	Intensity (I)	Relative intensity ($\frac{I}{I_1}$)
1	27.35	3.260	0.25	1.702
2	28.57	3.124	0.05	0.345
3	30.66	2.916	0.06	0.403
4	31.71	2.822	14.79	100.000
5	45.44	1.996	1.91	12.913
6	53.86	1.702	0.06	0.433
7	56.48	1.629	0.38	2.585

As shown in Table 1.2, the cubic system has the relationship among the interplanar spacing d, the Miller index and the lattice constant as follows:

$$\frac{1}{d_{hkl}^2} = \frac{h^2 + k^2 + l^2}{a^2}.$$ (1.25)

Therefore, the values of $\frac{1}{d_{hkl}^2}$ become simple integer ratios for each other. We use this principle.

Looking at the values of spacing $\frac{1}{d_{hkl}^2} \times 100$ in Table 1.6, the value of Peak 7 is four times that of Peak 1 and three times that of Peak 4. The value of Peak 5 is twice as large as that of Peak 4. Therefore, they are likely to be represented by simple integer ratios. When each of the $\frac{1}{d_{hkl}^2} \times 100$ values is divided by the d value (9.42) of Peak 1, the ratios are obtained as Ratio 1 column in this table. When the ratios in this column are tripled to obtain another series of the ratios in Ratio 2 column, all these ratios become simple integers without Peaks 2 and 3. It has been found by the detailed chemical analysis that Peaks 2 and 3 are originated from $MgCl_2$ as an impurity contained in NaCl. Therefore, excluding these peaks, further analysis is carried out. As can be seen from Ratio 2 column, the values become 3.00, 4.00, 8.00, 11.00 and 12.00. Therefore, the ratios are integers of 3, 4, 8, 11 and 12.

From Equation (1.25),

$$\frac{1}{d_{hkl}^2} = \frac{h^2 + k^2 + l^2}{a^2} \propto (h^2 + k^2 + l^2)$$

Table 1.6.   Indexation method from the spacings observed in the X-ray powder diffraction study for NaCl.

Peak No.	d(Å)	$\frac{1}{d^2} \times 100$	Ratio 1	Ratio 2	(hkl)	$a = d\sqrt{h^2 + k^2 + l^2}$
1	3.260	9.41	1.000	3.00	(1 1 1)	5.65
2	3.124	10.25	1.089	3.26×	—	—
3	2.916	11.76	1.250	3.75×	—	—
4	2.822	12.56	1.335	4.00	(2 0 0)	5.64
5	1.996	25.10	2.667	8.00	(2 2 0)	5.65
6	1.702	34.52	3.668	11.00	(3 1 1)	5.64
7	1.629	37.68	4.000	12.00	(2 2 2)	5.64

The simplest ratios   Indexation   Verification by lattice constant calculation

The $(h^2 + k^2 + l^2)$ values become these integers, when the (hkl) values are $(1\,1\,1)$, $(2\,0\,0)$, $(3\,1\,1)$ and $(2\,2\,2)$ respectively. This way of determination of the Miller indices for the diffraction peaks is called **indexing**.

However, novices will think that such an indexing method is truly like magic. Therefore, it should be certified by calculation of the lattice constant in the following method whether the indexing is correct or not.

Procedure 3

Calculate the lattice constant a of NaCl. If the indexing determined above is correct, the value of the lattice constant calculated from each diffraction line should be the same for any peak. Make sure of this.

From Equation (1.25), $a = d\sqrt{h^2 + k^2 + l^2}$.

Using this equation, calculate the cubic lattice constant a as shown in the rightmost column of Table 1.6. As can be seen from this column, the lattice constant is constant within the margin of error, which proves that this indexing is correct.

NaCl has a cubic lattice. However, in the cubic system, there are three Bravais lattices, a simple cubic lattice (P), a face-centered cubic lattice (F), and a body-centered cubic lattice (I), as can be seen from Fig. 1.7. Therefore, the question remains as to which of P, F and I the cubic lattice of NaCl is. In order to solve this problem, the number Z of molecules (atoms) in the unit cell is calculated as follows.

Procedure 4

The number Z of atoms in the unit cell is determined from the experimentally observed density of NaCl, $\rho_{obs} = 2.163\,\text{g/cm}^3$.

The density ρ is theoretically given as

$$\rho = \frac{MZ}{VN} \tag{1.26}$$

M: atomic weight (molecular weight)
Z: Number of atoms in unit cell (number of molecules)
V: unit cell volume
N: Avogadro constant $6.02 \times 10^{23} \, \mathrm{mol}^{-1}$

Therefore,

$$Z = \frac{\rho_{obs} V N}{M}$$
$$= 2.163 \, \mathrm{g/cm}^3 \times (5.65 \times 10^{-8} \, \mathrm{cm})^3$$
$$\times (6.02 \times 10^{23}/\mathrm{mol})/58.44 \, \mathrm{g/mol}$$
$$= 4.02$$

here,

For a simple cubic lattice (P), $Z = 1$.
For a face-centered cubic lattice (F), $Z = 2$.
For a body-centered cubic lattice (I), $Z = 4$.

Since the value from the lattice constant has been calculated as $Z = 4.02$, it can be concluded that the crystal of NaCl form a face-centered cubic lattice (fcc).

By the way, we can confirm the Z number also from Table 1.4. As can be seen from this table,

$$\mathrm{Na} \quad \frac{1}{8} \times 8 + \frac{1}{2} \times 6 = 4$$
$$\mathrm{Cl} \quad \frac{1}{4} \times 12 + 1 = 4$$

Thus, it can be confirmed that the crystal of NaCl contains four molecules in the unit cell.

Procedure 5

It can be concluded also from the extinction rule that the crystal of NaCl forms a face-centered cubic lattice (fcc). So, we consider it from Equation (1.24) and the conclusion in Section 1.12.2.

The Miller indices assigned above and the peak intensities in Fig. 1.24 are summarized below:

Peak No.	(hkl)	Intensity
1	(1 1 1)	Weak
4	(2 0 0)	Strong
5	(2 2 0)	Strong
6	(3 1 1)	Weak
7	(2 2 2)	Strong

As can be seen from these intensities, when h, k and l are all even, the intensities are strong; when h, k and l are all odd, the intensities are weak. It is consistent with the rules derived from Equation (1.24). Furthermore, when one is odd and the others are even, or when one is even and the others are odd, the reflections disappear. It holds the extinction rule of fcc. Thus, it can be proven also from the extinction rule that the crystal lattice of NaCl has a face-centered cubic lattice.

In the actual procedure of X-ray single crystal analysis, we will follow Procedure 1 to Procedure 3 and then Procedure 5 mentioned above, determine the space lattice from extinction rules, and finally calculate the theoretical density ρ_{calcd} in Procedure 4 to compare with the actually measured density ρ_{obs}. If $\rho_{calcd} = \rho_{obs}$, this proves that the analysis is correct.

References

[1] *Atkins Physical Chemistry*, 8th edition, Chapter 20, Tokyo Kagaku Dojin, Tokyo, 2009: 5th edition, Kittel, *Solid Physics Introduction*, Chapters 1 and 2, Maruzen, Tokyo, 1986.

[2] Toshio Sakurai, *Guide to X-ray Crystallography*, Shokabo, Tokyo, 2003.

[3] Kentaro Yano, Shigeru Ishihara, Fundamental Mathematics for Technologists, Yuhuabo, Tokyo, 1972.

[4] Ohashi Yuji, *Journal of the Crystallographic Society of Japan* 57, 131–133, 2015.

[5] Information brochure of Tokyo University of Agriculture and Technology, Former Textile Museum (now Science Museum), 2009.

Chapter 1. Exercises

1. Explain what white X-rays and characteristic X-rays are.
2. Geometrically prove Bragg's condition.
3. In an orthorhombic lattice, mark and show each plane of (110), (111), (210), (200), (310) represented by Miller index.
4. Calculate the spacing d_{hk0} of a two-dimensional rectangular lattice.
5. Calculate the spacing d_{hkl} of a three-dimensional orthorhombic lattice.
6. $SbCl_3$ is orthorhombic, and its lattice constants are $a = 8.12\,Å$, $b = 9.47\,Å$, $c = 6.37\,Å$. Calculate the spacing of the (411) plane.
7. Describe what you can reveal from single crystal X-ray diffraction and powder crystal X-ray diffraction, respectively.
8. Calculate the structure factor of NaCl, and obtain the extinction rule of the face-centered cubic lattice.
9. In X-ray diffraction, although both NaCl and KCl are face-centered cubic structures, why does KCl apparently look like a simple cubic structure? (cf. Question 8 and Moore Physical Chemistry p. 853)
10. Carry out the crystal structure analysis of NaCl using the X-ray powder pattern shown in Fig. 1.24 in the text. Follow procedures from 1 to 5 written below. (cf. Section (xiii) in the textbook.)

(Procedure 1) Using $\lambda = 1.5418\,Å$ of Cu Kα, calculate the interplanar spacing d of each diffraction line and the relative intensity ratio (I/I_1).

(Procedure 2) It is already known from microscopic observation that NaCl is cubic system. With this in mind, index each diffraction line.

(Procedure 3) Calculate the lattice constant of NaCl. If your indexing in Procedure 2 is correct, the value calculated from each diffraction line should be almost constant. Check your results in this way.

(Procedure 4) Determine the number of atoms in the unit cell using the observed density $d_{obs.} = 2.163\,g \cdot cm^{-3}$ of NaCl.

(Procedure 5) Examine your indexing from the extinction rule derived in Procedure 2, and prove that the NaCl crystal has a face-centered cubic lattice.

11. Draw a two-dimensional reciprocal lattice plane using the lattice constant of NaCl, $a = 5.65\,\text{Å}$, and show that the (200) and (220) planes indexed in Question 10, exist just on the lattice points (intersections) in this plane. (cf. Fig. 1.25 as a solution example)

12. The crystal of $NiSO_4$ is orthorhombic. The lattice constants are $a = 6.34\,\text{Å}$, $b = 7.84\,\text{Å}$, $c = 5.16\,\text{Å}$, and the observed density of the crystal was $3.9\,\text{g·cm}^{-3}$. Using these values, calculate the number (Z) of molecules included in the unit cell. From this Z value, decide which orthorhombic lattice it has among four lattices of the simple orthorhombic (P), C-centered

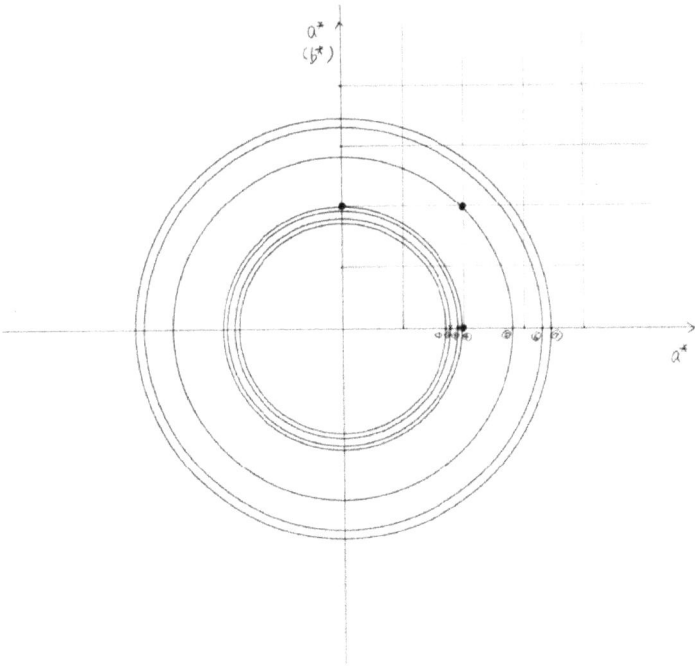

Figure 1.25. DebyeScherrer rings calculated from the powder pattern in Fig. 1.24.

orthorhombic (C), face-centered orthorhombic (F) and body-centered orthorhombic (I) shown in Fig. 1.7.

13. Calculate the filling factor for the following unit cells. (a) simple cubic lattice, (b) body-centered cubic lattice (bcc), (c) face-centered cubic lattice (fcc).

Chapter 2

Polymorphism

2.1 Polymorph and Polymorphism [1–3]

Polymorphism refers to the phenomenon in which any compound has plural crystalline forms. When the crystalline forms are different even in the same compound, their physical properties such as density, hardness, optical and electrical properties, solubility, medical efficacy etc., are completely different. It is essential to understand polymorphism for scientists seeking high functionality, pharmaceutical scientists seeking high medical efficacy, and students aiming for them.

Below are some examples of polymorphism.

Example 1. Carbon (C_∞) (cf. Fig. 2.1)

- Diamond: cubic system $a = 3.567\,\text{Å}$, hardness 10, generally insulator $\rho = 10^{13}\,\Omega\,\text{cm}$. cf. Fig. 2.1[A].
- Graphite: hexagonal system $a = 2.456\,\text{Å}$, $c = 6.696\,\text{Å}$, hardness 1 to 2, semimetal $\rho = 10^{-3}\,\Omega\,\text{cm}$. cf. Fig. 2.1[B].

Thus, although the molecules of both diamond and graphite are represented as C_∞, diamond has a hardness of 10, while graphite has a hardness of 1 to 2; diamond is an insulator, while graphite is a semimetal. Hence, physical properties are completely different depending on crystal polymorphism.

Since the molecules of both diamond and since graphite are represented by the same C_∞ and they have only different crystal structures, they are crystal polymorphs. However, C_{60} and C_{70}

[A]

[B]

[C]

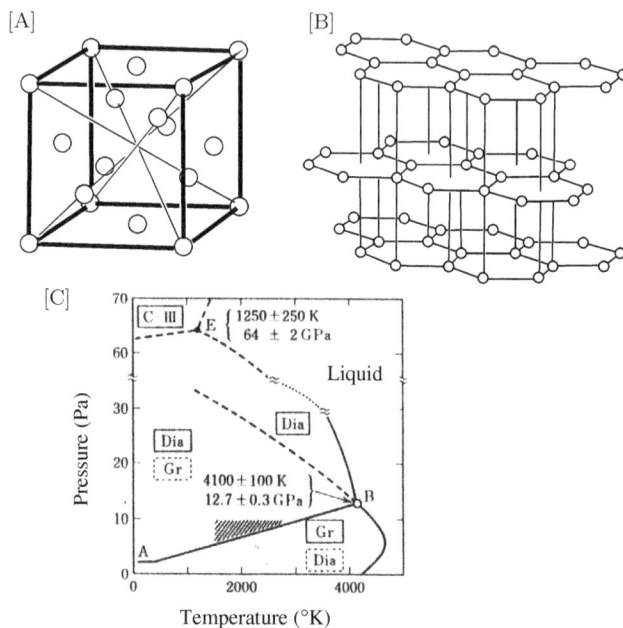

Figure 2.1.   Diamond and graphite.

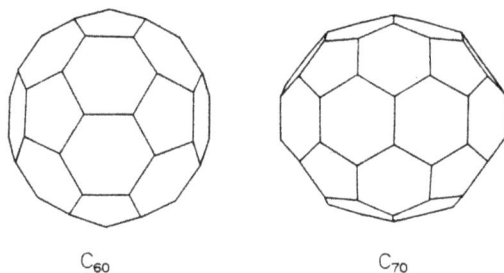

C_{60} C_{70}

Figure 2.2. C_{60} and C_{70}.

(Fig. 2.2) have different molecular formulae from C_{∞}, so they are not called as polymorphs of C_{∞}. It should be remarked that C_{60} and C_{70} are allotropes of C_{∞} but not polymorphs of C_{∞}.

Example 2. Ice (H_2O) (cf. Fig. 2.3, Table 2.1)
As you can see in Fig. 2.3 and Table 2.1, up to now 17 crystal polymorphs of ice (H_2O) have been discovered. The ice that we see

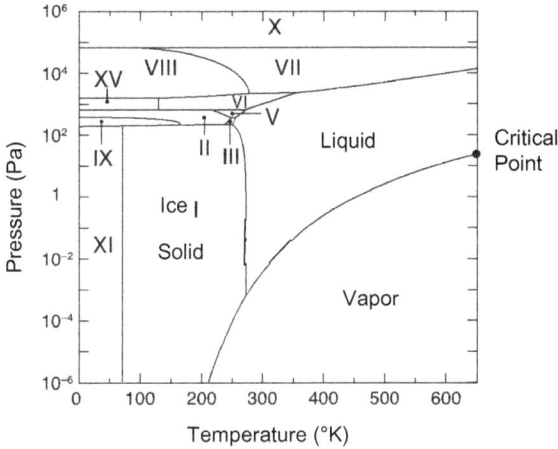

Figure 2.3.   Phase diagram of water.

Table 2.1.   The ice polymorphs.

Polymorph	Crystal system
I	Cubic
II	Rhombohedral
III	Tetragonal
IV	Rhombohedral (metastable)
V	Monoclinic
VI	Tetragonal
VII	Cubic
VIII	Tetragonal
IX	Tetragonal
X	Cubic
XI	Orthorhombic
XII	Tetragonal (metastable)
XIII	Monoclinic
XIV	Orthorhombic
XV	Pseudo-orthorhombic
XVI	Cubic
XVII	Hexagonal

in our daily life is ice I (Ice I), which is a polymorph that floats on water. On the other hand, ice polymorphs (III, V and VII) sink in water. This can be resulted from the derivative $\frac{dP}{dT}$ of the solid–liquid boundary in the P-T phase diagram of Fig. 2.3 that the derivative is negative for ice I and positive for ice III, V, and VII (for details, see Section 3.2) Moreover, it should be noted that there are polymorphs I, VII, X and XVI having different crystal structures even in the same cubic system.

Example 3. Sulfur (crown-like S_8 molecule) (cf. Figs. 2.4 and 2.5)

- $S\alpha$: orthorhombic sulfur, melting point = 112°C.
- $S\beta$: monoclinic sulfur, melting point = 119°C

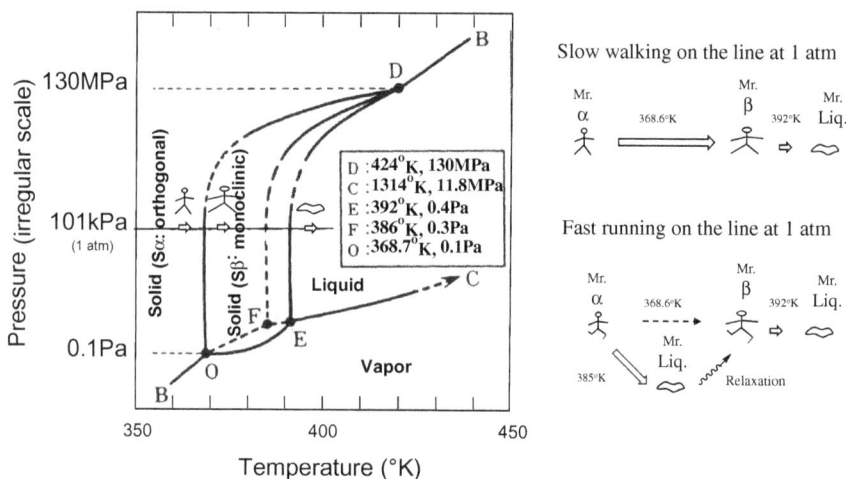

Figure 2.4.   Phase diagram of sulfur.

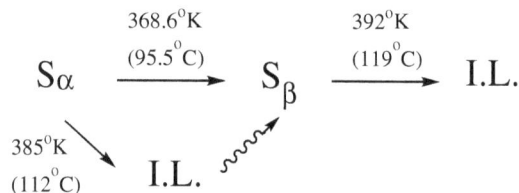

Figure 2.5.   Double melting behavior of sulfur.

Sulfur is a yellow element commonly found in volcanic fumaroles. Though they are commonly found in rocks, as can be seen from Fig. 2.5, their melting points are strangely low. That is because sulfur forms a crown-like S_8 molecule, its structure closely resembles that of an organic substance. There are two crystal polymorphs of orthorhombic sulfur ($S\alpha$) and monoclinic sulfur ($S\beta$) for the crown-like S_8 molecule, and their melting points are $112°C$ and $119°C$, respectively. Sulfur shows **double melting behavior** as shown in Fig. 2.5 (for details, see Section 3.1).

Example 4. Medicine

Sometimes there are more than 10 crystal polymorphs in one medicine. Solubility varies depending on polymorphic forms present, and there are high efficacy drugs, low efficacy drugs, rapid efficacy drugs, and slow efficacy drugs. Therefore, polymorphism is an important issue in pharmaceutics.

Other examples. Soap, paraffin, etc.

As can be seen from the above examples, polymorphism is crystalline phases of a compound that result from at least two different molecular arrangements in the crystalline state. The molecular structure does not change between the two polymorphs. However, there are partial rotation and small distortion of the bond.

The two polymorphs differ in crystal structure but are identical in the liquid and vapor states. Therefore, there is generally no polymorphism in liquids and vapors. As an exception, it is well-known that helium is present in polymorphic forms of liquids: liquid He I and He II (cf. Fig. 2.6). Recently, a liquid-water phase transition has been discovered: high density water and low density water [4]. However, no polymorphs of liquids have been found other than helium and water. Thus, one may generally think that "polymorphs exist in the solid phase but not in the liquid phase."

Also, as you can see in Fig. 2.7, it should be noted that geometric isomers and tautomers are not polymorphic because each melt (liquid phase) gives a different liquid.

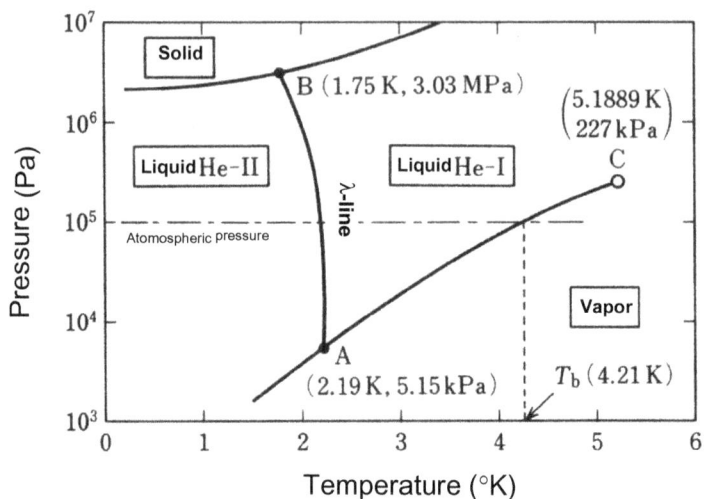

Figure 2.6.   Phase diagram of He.

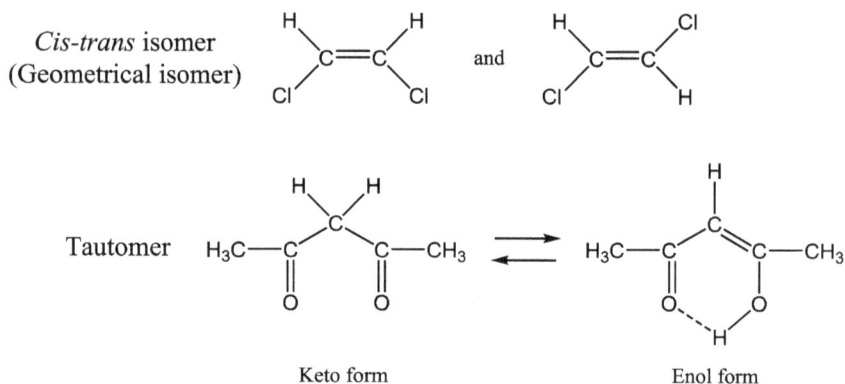

Figure 2.7.   Cis-trans isomer and tautomer.

2.2   First-order Phase Transition and Second-order Phase Transition

Before describing the first-order phase transition and the second-order phase transition, the $G(\mu)$–T diagram will be described at first. In order to fully understand these phase transitions, it is essential to understand the G–T diagrams.

2.2.1 *G(μ)–T diagram*

G: Gibbs energy, H: Enthalpy, T: Temperature, S: Entropy,

$$G = H - TS \tag{2.1}$$

$$\therefore \quad \left(\frac{\partial G}{\partial T}\right)_p = -S \tag{2.2}$$

For pure substances, $G = \mu$ (μ: chemical potential),

$$\left(\frac{\partial \mu}{\partial T}\right)_p = -S \tag{2.2$'$}$$

Since it is always S > 0, in the G(μ)–T diagram, the slope $\left(\frac{\partial G}{\partial T}\right)_p$ of each straight line is always negative and always downward to the right. It is necessary to pay close attention that when you would draw a line upward to the right, it would be a violation of the law of entropy increase (the second law of thermodynamics). Figure 2.8 shows a schematic G–T diagram of crystal (K), liquid (L) and vapor (V).

2.2.2 *What the G–T diagram tells*

(1) Usually, when the temperature is applied to the substance, the vapor phase steeply increases the degree of randomness (entropy

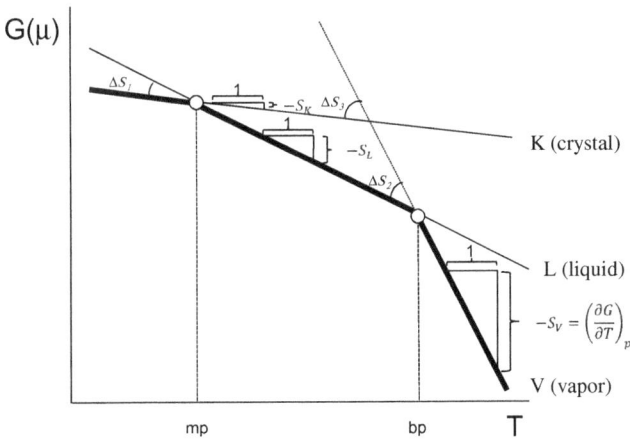

Figure 2.8. Meaning of G–T diagram.

S) most rapidly, so the slope of the straight line of vapor is the largest in the G–T diagram (Fig. 2.8). Therefore,

$$S_K < S_L < S_V \tag{2.3}$$

(2) Since it can be considered as $\Delta S \propto$ (angle between two straight lines), in Fig. 2.8

$$\Delta S_1 < \Delta S_2 < \Delta S_3 \tag{2.4}$$

As can be seen from this figure, ΔS_3 is the largest in the angles. It is because the order difference between the crystalline phase and the vapor phase is the largest.

(3) The phase showing the lowest free energy G at a certain temperature T is the most stable.

$$
\begin{aligned}
\text{m.p.} > \text{T} & \qquad G_K < G_L < G_V \\
\text{m.p.} < \text{T} < \text{b.p.} & \qquad G_L < G_K, G_V \\
\text{T} > \text{b.p.} & \qquad G_V < G_L < G_K
\end{aligned}
\tag{2.5}
$$

Therefore, the phases remarking with thick line in Fig. 2.8 are the most stable at that temperature.

(4) When looking at only the most stable phases, the phase transitions can be occurred as follows:

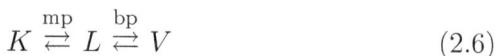

$$K \overset{\text{mp}}{\rightleftarrows} L \overset{\text{bp}}{\rightleftarrows} V \tag{2.6}$$

Note 1: This is when the phase transition rate is not taken into consideration. When rapid heating or rapid cooling, other phenomena such as **double melting behavior** may appear.

Note 2: Only the most stable phase does not always exist under the temperature and pressure. The metastable phases can exist at the same temperature and pressure. Example: under ambient temperature and pressure, both diamond (metastable phase) and graphite (most stable phase) can coexist. If only the most stable phase could exist under ambient temperature and pressure, no diamond could exist under ambient temperature and pressure. In reality, you can buy diamonds as jewels. Diamond does not change to graphite under ambient temperature and pressure.

This is because there is a large barrier of activation energy for the phase transition.

(5) When $\Delta G = 0$, a phase transition occurs.

$$\left.\begin{array}{ll} \text{At mp,} & \Delta G = G_K - G_L = 0 \\ \text{At bp,} & \Delta G = G_L - G_V = 0 \end{array}\right\} \tag{2.7}$$

Thus, at mp a phase transition between crystal and liquid occurs, and at bp, a phase transition between liquid and vapor occurs. As can be seen from these examples, the phase transition occurs at the intersection on the G–T diagram. At this point of intersection, the two phases are at the same Gibbs energy, and since $\Delta G = 0$, a phase transition occurs.

However, the speed of phase transition is a matter of kinetics, and there is an activation energy barrier that can not be represented in this G–T diagram.

2.2.3 *Difference between first-order phase transition and second-order phase transition*

As above, after understanding the nature of the G–T diagram, we will look at the difference between the first-order phase transition and the second-order phase transition. The G–T diagrams of the first-order phase transition and the second-order phase transition are schematically drawn in Figs. 2.9[A] and 2.9[B], respectively.

For the first-order phase transition (Fig. 2.9[A]), the stability of the phase is represented by a straight line in the G–T diagram. The straight line I has a gentle slope, while the straight line II is a steep slope. When heating from the low temperature side, the two straight lines intersect at the temperature T_t, so the phase transition from the phase I to the phase II occurs at the temperature T_t. Therefore, as the heating progresses, the slope suddenly increases at the phase transition temperature T_t. From Equation (2.2), the slope of the straight line corresponds to the first derivative $\left(\frac{\partial G}{\partial T}\right)_p (= -S)$. Therefore, the entropy S jumps discontinuously at this temperature T_t. To see it more clearly, an S–T diagram is drawn below the

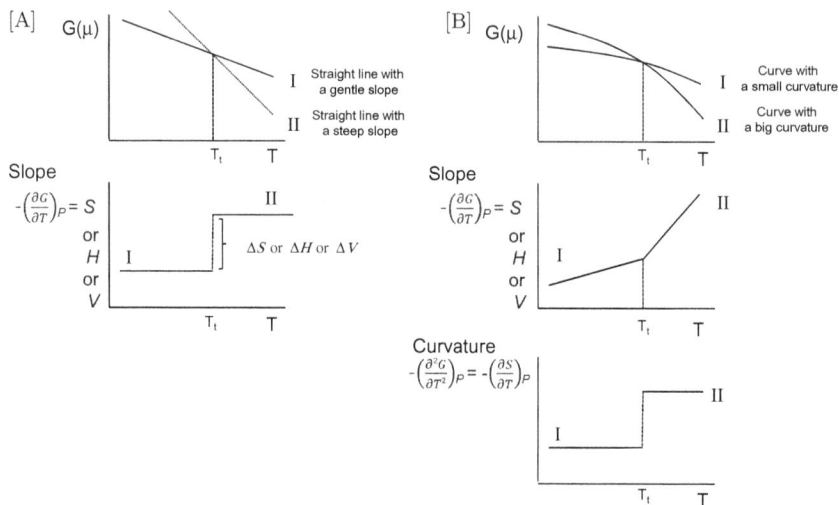

Figure 2.9. [A] 1st order phase transition and [B] 2nd order phase transition.

G–T diagram. As can be seen from this S–T diagram, at T_t the lower horizontal line I jumps discontinuously to the higher horizontal line II. Moreover, at the phase transition temperature T_t of the first-order phase transition, the enthalpy H and the volume V also jump in this manner, as well as the entropy S (see thermodynamics for the detailed derivation).

For the second-order phase transition (Fig. 2.9[B]), the stability of the phase is represented by a curve in the G–T diagram. The curve I has a gentle curvature, while the curve II has a steep curvature. These two curves cross at the temperature T_t. Therefore, at this temperature T_t a phase transition takes place from the phase I to the phase II. At T_t, the curvature (acceleration) changes from a small one to a big one. To see it more clearly, the S–T diagram is drawn under the G–T diagram. As can be seen from this S–T diagram, at T_t, the straight line I having a gentle slope changes to the straight line II having a steep slope. Furthermore, a diagram of the slope of the straight line, namely, the second derivative $-(\partial^2 G/\partial T^2)_P (= $ curvature $=$ acceleration) vs. temperature is drawn below the S–T diagram. As can be seen from this $-(\partial^2 G/\partial T^2)_P$–T diagram, the lower horizontal line I jumps discontinuously to the higher horizontal line II at the T_t temperature.

According to the characteristic G–T diagrams mentioned above, the difference between the first-order phase transition and the second-order phase transition can be easily understood as follows.

The first-order phase transition: the first-order derivative of Gibbs energy (chemical potential) is discontinuous.

(A phase transition in which V, H, S jump discontinuously is a first-order phase transition.)

The second-order phase transition: the second-order derivative of Gibbs energy (chemical potential) is discontinuous.

(A phase transition in which V, H, S do not jump discontinuously is a second-order phase transition.)

Note: V, H, S is a word analogy of Video Home System. It makes us to easily remember these characteristics.

2.2.4 *Concerning* ΔV

In the case of the first-order phase transition, when a crystal-crystal phase transition is observed using a polarization microscope with a temperature controller, rapid swelling and wrinkle formation can be observed at the phase transition temperature during heating; rapid cracks and shrinkage can be seen at the phase transition temperature during cooling. Since these volume changes ΔV occur suddenly at certain temperatures, it is possible to determine the first order phase transition temperature from microscopic observation. On the other hand, in the second-order phase transition, these changes occur gradually, so it is not clear where the second-order phase transition occurs. Therefore, you cannot determine the second order phase transition temperature from microscopic observation.

2.2.5 *The cause of the second order phase transition*

There are two major causes of the second order phase transitions. (For details, see, for example, Seki *et al.*, *"Physical chemistry of Pure Substances"* pp. 194 – 197 [5].)

2.2.5.1 *Order–disorder transition*

(Example) KCN

KCN is an ion crystal which can be represented as $K^+(CN)^-$.

At low temperatures, all the anions $(CN)^-$ are either CN^- or NC^- oriented.

Therefore, since the number of states is only one, the entropy S_l at this time is

$$S_l = k \ln 1 = 0$$

At high temperature, CN direction: NC direction $= 1 : e^{-\Delta E/kT}$

Above the transition temperature ($\Delta E = 0$), the numbers of both directions are equal.

Therefore, since there are 2^N ways of states, the entropy S_h at this time is

$$S_h = k \ln(2^N) = Nk \ln 2 = R \ln 2$$

$$\therefore \ \Delta S = S_h - S_l = R \ln 2 = 1.38 \text{ cal deg}^{-1}\text{mol}^{-1} \quad \text{(calculated value)}$$

$$\text{Measured value } \Delta S = 1.32 \text{ cal deg}^{-1}\text{mol}^{-1}$$

The calculated value matches with the theoretical value of the entropy. Thus, it was theoretically proved that this second-order phase transition is originated from an order-disorder of the direction of CN ions.

2.2.5.2 *Rotational transition*

The nearly spherical molecules CH_4, CCl_4, C_{60} undergo a second-order phase transition from crystals to plastic crystals. This second order phase transition is known as rotational phase transition. For example, crystals of CCl_4 form a body-centered cubic lattice, and the orientation of each molecule is regularly arranged at low temperatures, but as heating, the molecules rotate while forming a cubic lattice, and only the orientation order of the molecules gradually disappear. When the order of the posture completely disappears, it becomes a plastic crystal phase. This is because the rotational barrier is small when the molecules are close to spherical.

This rotational transition is also similar to the order-disorder transition above-mentioned.

2.3 Phase Diagram

There are three intensive variables that determine the chemical potential μ of the system (Gibs free energy G in the case of pure substances): pressure (P), temperature (T), and component (X). Therefore, it is P, T and X that determine the phase of the system. Accordingly, there are several types of phase diagrams depending on the combination of intensive variables, which can be used according to the purpose.

For example, when looking at the stability of crystal polymorphs, Gibbss free energy G (or chemical potential μ) vs. temperature phase diagram, G (μ)–T diagram, is often used. In addition, when looking at the phase of pure substance (when X = 1), a pressure-temperature phase diagram, P–T diagram is often used. The P–T diagram may be simply called as a phase diagram or a state diagram. When looking at the phase of a two-component mixture system, a temperature-composition diagram (T–X diagram) under a constant pressure, or a pressure-composition diagram (P–X diagram) at a constant pressure is often used. These T–X and P–X figures are often employed in the cases of purification of substances by distillation and recrystallization.

In the followings, each of these phase diagrams will be used in order to consider crystal polymorphs.

2.3.1 *G–T diagram*

2.3.1.1 *Relationship between polymorphisms in the*
G–T diagram: Enantiotropic relationship
and monotropic relationship

The nature of the G–T diagram has been already described in Section 2.2.1. Hereupon, by using G–T diagrams, the differences in phase transition are explained in both the cases of an enantiotropic relationship (Fig. 2.10[A]) and a monotropic relationship (Fig. 2.10[B]) for two crystal polymorphs I and II.

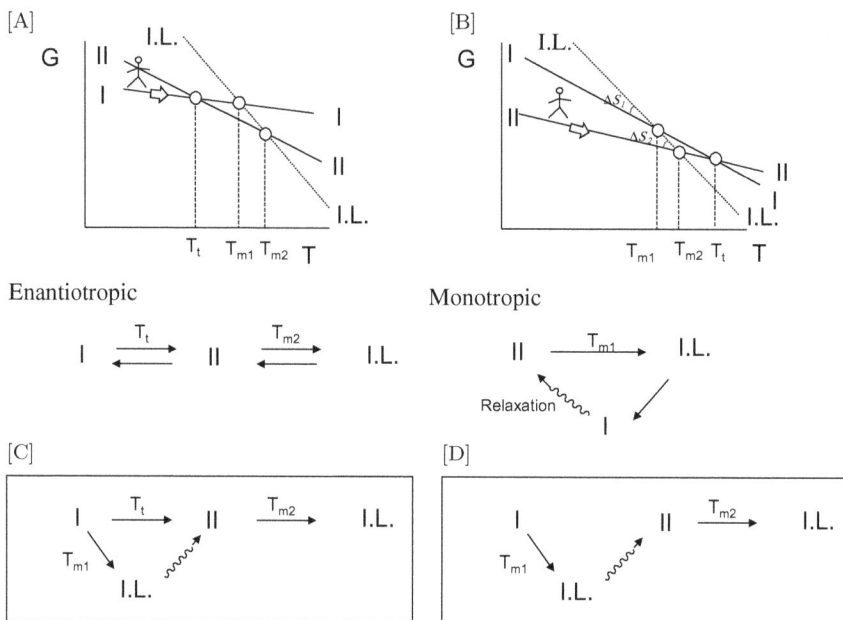

Figure 2.10. Relationship between two crystalline polymorphs I and II: [A] enantiotropic relation and [B] monotropic relation; [C] and [D] double melting behaviors corresponding to these relationships [A] and [B], respectively.

First, in the G–T diagram of Fig. 2.10[A], when the crystal polymorph I is heated from a low temperature sufficiently slowly, you go along the straight line I to the temperature T_t. At the temperature T_t, you have a complete phase transition from polymorph I to polymorph II, and then you go along line II. When you reach at the temperature Tm2, you lose your form of crystal polymorph II and are completely transformed into an amorphous isotropic liquid (IL). Over the temperature Tm2, you go along the straight line IL. If this liquid IL is cooled sufficiently slowly, you are completely transformed from IL to the crystal polymorph II at Tm2, and on further cooled, you are completely transformed from polymorph II to polymorph I at the temperature T_t. Thus, you return to the first polymorph I along the original route. Therefore, the same path is taken in the heating process and the cooling process. The phase transition sequence is summarized below the G–T diagram in Fig. 2.10[A]. As can be seen from this sequence, the crystal polymorphs I and II undergo a phase transition of mirror image symmetry in temperature. In the case, it is termed that crystal polymorphs I and II have an enantiotropic relationship.

Next, in the G–T diagram of Fig. 2.10[B], when the crystal polymorph II is heated from a low temperature, you go along the straight line II to the temperature Tm2. At the temperature Tm2, you lose your form of crystal polymorph II and are completely transformed into an amorphous isotropic liquid (IL). Over the temperature Tm2, you go along the straight line IL. When cooled this hot liquid IL, you travel in the reverse direction along the original straight line IL. Even if you reach at Tm2, you do not transform into the crystal polymorph II from the isotropic liquid IL and overrun to reach Tm1 by supercooling. Here you transform into crystal polymorph I. This is because Tm1 and Tm2 are very close to each other. Furthermore, since $\Delta S_1 < \Delta S_2$, the order difference (ΔS_1) between the isotropic liquid IL and the crystal polymorph I is smaller than that (ΔS_2) between the isotropic liquid IL and the crystal polymorph II. This means that the randomness degree of the crystal polymorph I is closer to the isotropic liquid than the crystal polymorph II. Therefore, the transition from IL to I is more likely (cf. Property (2) in Section 2.2.2. As can be seen from the G–T diagram in Fig. 2.10[B], the crystal polymorph I thus produced is unstable below the temperature Tm1, because it is above crystal polymorph II. Therefore, some accidental stimulation may cause a relaxation from crystal polymorph I to crystal polymorph II. By this relaxation, you return to the initial crystal polymorph II. Thus, the paths are different between the heating process and the cooling process. The phase transition sequence is summarized below the G–T diagram in Fig. 2.10[B]. As can be seen from this sequence, crystal polymorphs I and II undergo the thermally asymmetric phase transitions. In the case, it is termed that crystal polymorphs I and II have a monotropic relationship.

When crystal polymorphs I and II have an enantiotropic relationship, $T_t <$ Tm1, Tm2 holds, as can be easily understood from the G–T diagram of Fig. 2.10[A]. Hence, the crystal–crystal phase transition temperature T_t is lower than the melting points Tm1 and Tm2 of these polymorphs.

On the other hand, when the crystal polymorphs I and II have a monotropic relationship, $T_t > $ Tm1, Tm2 holds, as can be easily understood from the G–T diagram of Fig. 2.10[B]. Hence, the crystal–crystal phase transition temperature T_t is higher than the melting points Tm1 and Tm2 of these polymorphs. The crystal–crystal phase transition temperature T_t is a virtual phase transition temperature that cannot be observed in practice.

2.3.1.2 *Double melting behavior [1–3]*

In the above description, in the G–T diagram of Fig. 2.10[A], the heating speed of the crystal polymorph I was limited to the case of sufficiently slow heating from a low temperature. In this case of slow heating, the melting point Tm1 of the crystal polymorph I does not appear. However, when the crystal polymorph I is heated rapidly, the melting point Tm1 of the crystal polymorph I can be observed. Figure 2.10[C] shows the phase transition sequence when the crystal polymorph I is heated rapidly. When heated rapidly, crystal polymorph I does not completely convert to crystal polymorph II and you overrun through the crystal–crystal phase transition temperature T_t, and you reach to the melting point Tm1 of crystal polymorph I. At Tm1, it melts into the isotropic liquid IL. Since this IL is higher than the crystal polymorph II in the G–T diagram, it is more unstable than polymorph II, as can be seen from Fig. 2.10[A]. Therefore, the IL is rapidly relaxed to the crystal polymorph II by using the seeds of the crystal polymorph II partially generated in the crystal-crystal phase transition. When the resulted crystal polymorph II is further heated, it melts once more into the isotropic liquid IL at the melting point Tm2 of the crystal polymorph II. Thus, when the crystal polymorph I is heated rapidly, a double melting behavior can be observed. Such a phase transition sequence is illustrated in Fig. 2.10[C]. In this sequence, the straight arrows and the wavy arrow represent phase transitions and relaxation, respectively. The faster the heating rate becomes, the more clearly this double melting behavior appears. This is because the crystal polymorph I has more time to completely transform into the crystal polymorph II over T_t, when the crystal polymorph I is heated sufficiently slowly. In this case, the crystal polymorph I completely transforms into the crystal polymorph II, so that the melting point Tm1 can not be observed.

Next, we will consider the case that the crystal polymorphs I and II have a monotropic relationship shown in the GT diagram in Fig. 2.10[B]. When the crystal polymorph I is heated from a low temperature without the relaxation from crystal polymorph I to crystal polymorph II, the crystal polymorph I is heated along line I to reach the melting point Tm1 of crystal polymorph I. Here all the crystals of polymorph I melt into the isotropic liquid IL. This IL is above the crystal polymorph II, as shown in the GT diagram of Fig. 2.10[B], so that the IL is more unstable than the crystal polymorph II. Hence, the IL is slowly relaxed into the crystal polymorph II and all the parts very slowly resolidify into polymorph II. When the crystal polymorph II resolidified here is further heated, it melts once more into the isotropic liquid IL at the melting point Tm2 of the crystal polymorph II. Therefore, the double melting behavior is observed even if the crystal polymorphs I and II have a monotropic relationship. However, in this case, the slower the heating rate becomes, the more clearly the double melting behavior appears. This resolidification occurs very slowly because there is no seed of crystal polymorph II. If the heating rate is too fast, the IL reach over the melting point of crystal polymorph II (Tm2) before the resolidification occurs. This double melting behavior in the monotropic relationship is illustrated as a phase transition sequence in Fig. 2.10[D].

As described above, the appearance of these double melting behaviors depends on the heating rate both for the enantiotropic relationship and the monotropic relationship in crystal polymorphs I and II. The dependence on heating rate is quite opposite. Hence, the phase transition behavior changes depending on the heating rate. We should carefully pay attention on the heating rate when we observe the phase transition.

Note that double melting behavior does not occur on cooling stage. Because relaxation only occurs from the higher position to the lower position in the G–T diagram, it does not occur from the lower position to the higher position. In other words, relaxation only occurs from the unstable phase to the stable phase, but does not occur from the stable phase to the unstable phase. The reverse relaxation violates thermodynamic.

2.3.1.3   *DSC thermograms of double melting behavior [6]*

As mentioned above, the double melting behavior can be divided into two types (Figs. 2.10[A] and 2.10[B]). One type is observed when crystal polymorphs I and II has an enantiotropic relationship (the case in Fig. 2.10[A]). Another type is observed when crystal polymorphs I and II has a monotropic relationship (the case in Fig. 2.10[B]). We can clearly distinguish these two types if we observe the heating rate dependence of the differential scanning calorimeter (DSC) thermograms.

As can be seen from Fig. 2.11[A], the DSC thermograms show three endothermic peaks (peaks 1, 2 and 4) and one exothermic peak (peak 3). The endothermic peak 1 corresponds to the crystal–crystal

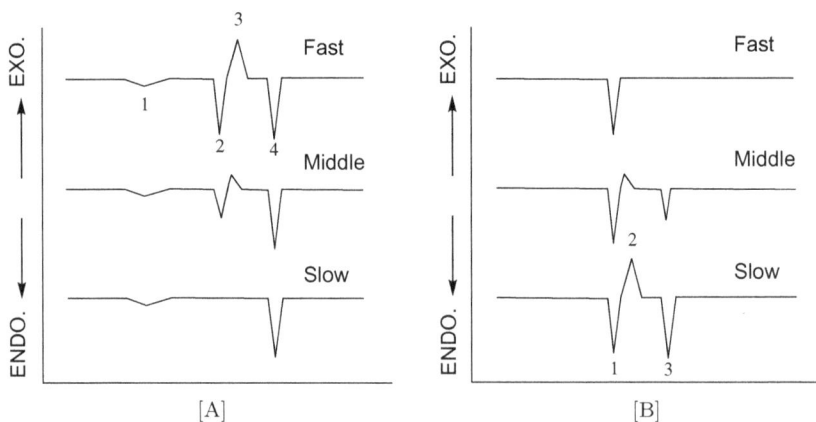

Figure 2.11.   Schematic representations of the heating rate dependence on DSC thermograms for [A] an enatiotropic relationship between Polymorphs I and II, and [B] a monotiotropic relation between Polymorphs I and II.

phase transition from polymorph I to II. This peak 1 is generally observed as a very broad peak. The endothermic peaks 2 and 4 correspond to the melting of polymorphs I and II, respectively. The exothermic peak 3 between peaks 2 and 4 corresponds to the recrystallization where the isotropic liquid (IL) resulted from the melting of polymorph I relaxes to polymorph II. As can be seen also from these thermograms in Type [A], the faster the heating rate becomes, the bigger the ratio of peak 2 (due to the melting of polymorph I) to peak 4 (due to the melting of polymorph II) becomes. This is because the faster the heating rate becomes, the more the amount of polymorph I remains without undergoing the crystal–crystal phase transition from I to II. Therefore, as can be seen in Fig. 2.11[A], the faster the heating rate becomes, the more clearly the double melting behavior can be observed.

When polymorphs I and II has a monotropic relationship between (in the case of Fig. 2.10[B]), the DSC thermograms show two endothermic peaks (peaks 1 and 3) and one exothermic peak (peak 2) as can be seen from Fig. 2.11[B]. These endothermic peaks 1 and 3 correspond to the melting of polymorphs I and II, respectively. The exothermic peak 2 between peaks 1 and 3 corresponds to the recrystallization that the isotropic liquid (IL) resulted from the melting of polymorph I relaxes to polymorph II. The slower the heating rate becomes, the bigger the ratio of peak 3 (due to the melting of polymorph II) to peak 1 (due to the melting of polymorph I) becomes. This is because the slower the heating rate, the more the recrystallization from the isotropic liquid IL to polymorph II occurs for the longer period. Accordingly, the slower the heating rate becomes the more clearly the double melting behavior can be observed, as can be seen from Fig. 2.11[B].

Thus, in Case A, the faster the heating rate becomes, the more clearly the double melting behavior can be observed. On the other hand, in Case B, the slower the heating rate becomes, the more clearly the double melting behavior can be observed. The heating rate dependence of cases A and B is quite opposite. Therefore, we can clearly distinguish these two types if we observe the heating rate dependence of the DSC thermograms.

2.3.1.4 *Relaxation*

The features of relaxation are as follows.

- Relaxation corresponds to the falling from metastable phase to more stable phase in the G–T diagram.
- Relaxation is not a phase transition.
- The relaxation temperature is indefinite but the phase transition temperature is constant at atmospheric pressure.
- Whether the relaxation is endothermic or exothermic is dependent on the difference between the two slopes of straight lines ($\propto \Delta S$) for the polymorphs I and II in the G–T diagram, when their melting points Tm1 and Tm2 are very close.

The fourth feature can be proved as follows. Figure 2.12 schematically illustrates when the slopes ($-S_1$ and $-S_2$) of crystal polymorphs 1(K_1) and 2(K_2) are in the cases of [A] $S_1 < S_2$ and [B] $S_1 > S_2$, respectively. We will consider these cases.

(Proof)

From the above Equation 2.1

$$G = H - TS$$
$$\therefore S = \frac{H-G}{T}$$

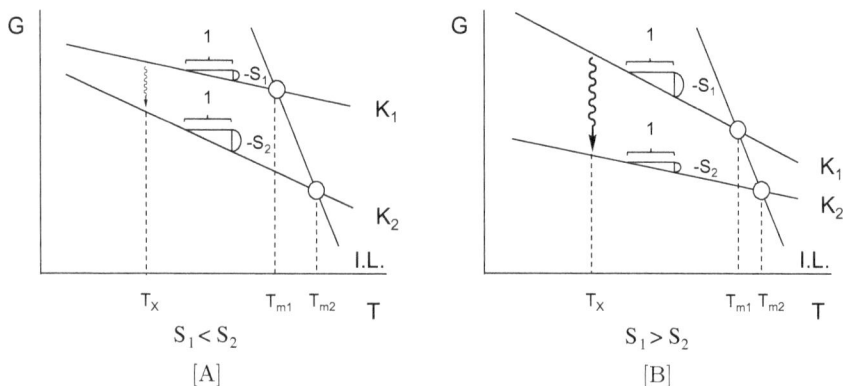

Figure 2.12.   G–T diagrams for [A] endothermic relaxation and [B] exothermic relaxation.

When relaxing with temperature T_X,

$$S_1 = \frac{H_1 - G_1}{T_X}, \quad S_2 = \frac{H_2 - G_2}{T_X}$$

$$\therefore \Delta S = S_2 - S_1 = \frac{H_2 - G_2}{T_X} - \frac{H_1 - G_1}{T_X}$$

$$= \frac{(H_2 - H_1) - (G_2 - G_1)}{T_X}$$

$$= \frac{\Delta H - \Delta G}{T_X}$$

(Case A) When $S_1 < S_2$,

$$\Delta S = S_2 - S_1 > 0$$

$$\therefore \Delta H - \Delta G > 0$$

$$\therefore \Delta H > \Delta G$$

Here, the relaxation from the straight line K_1 to the straight line K_2 is likely to occur when these lines are very close. In other words, relaxation is likely to occur when Tm1 and Tm2 are very close (Tm1 $\approx$ Tm2), as shown in Fig. 2.12. Therefore, in such a case, you can regard as

$$\Delta G \approx 0$$

$$\therefore \Delta H > 0$$

This means that endothermic relaxation occurs(in the rare cases).

(Case B) When $S_1 > S_2$,

$$\Delta S = S_2 - S_1 < 0$$

$$\therefore \Delta H - \Delta G < 0$$

$$\therefore \Delta H < \Delta G$$

From the same reason mentioned in Case A, you can regard as

$$\Delta G \approx 0$$

$$\therefore \Delta H < 0$$

This means that exothermic relaxation occurs (in the most cases).

(End of proof)

Accordingly, it is very interesting that only when you simply look at the difference in the slopes ($\propto \Delta$S) of the straight lines of the two polymorphs in the G–T diagram, you can intuitively predict whether the relaxation is endothermic or exothermic. Most of the relaxations are **exothermic relaxations** [Case B], but in the rare cases you can observe **endothermic relaxations** [Case A]. Therefore, we need very careful observation of phase transitions [7].

2.3.2   *P–T diagram*

2.3.2.1   *Consideration of the phase transition in the schematic P–T diagram*

Figure 2.13 shows a typical pressure vs. temperature phase diagram (P–T diagram) for a pure compound (X = 1).

Curves OA, OB and OC represent phase boundaries. On each of the curves, two phases coexist. Each of the phase boundaries can be also obtained by calculations using Clausius–Clapeyron Equation (for details, refer to the textbook of physical chemistry).

Point O represents a triple point, at which three phases of solid, liquid and vapor simultaneously coexist in equilibrium. Exceptionally, He has not solid–liquid–vapor triple point but two triple points of A (solid–liquid He I–liquid He II) and B (liquid He I–liquid He II–vapor), as already shown in Fig. 2.6.

In the P–T diagram of Fig. 2.13, when the temperature and the pressure are higher than both of the critical temperature (Tc) and the critical pressure (Pc), the interface between liquid and vapor disappears, resulting in a supercritical fluid indistinguishable the liquid and the vapor. Very interestingly, Nishikawa *et al.* found that the boundary line CD (dotted line) of density fluctuation still continues on the extrapolation of OC curve, and the region of the supercritical fluid can be divided into two areas [8].

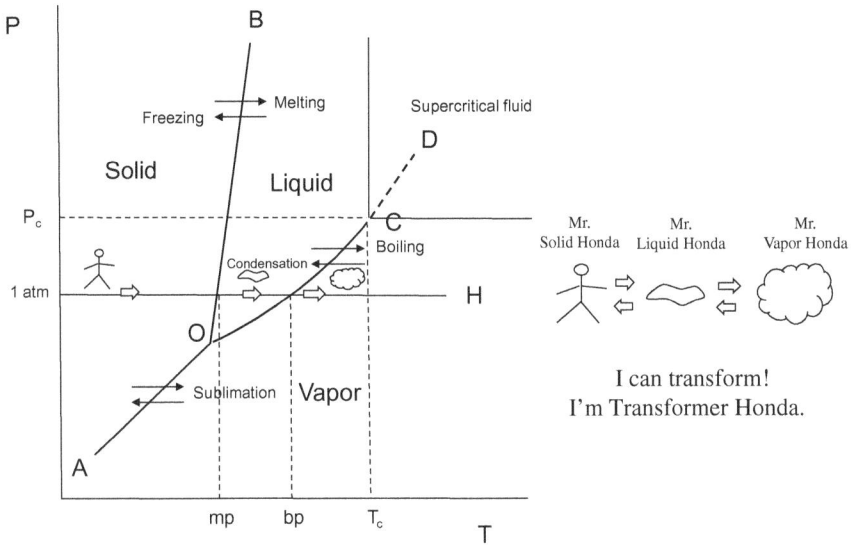

Figure 2.13. Schematic pressure-temperature (P–T) diagram of one component system. Curve OA: sublimation curve. Curve OB: melting curve. Curve OC: boiling curve or vapor pressure curve. Point O: triple point. Point C: critical point. Tc: critical temperature; Pc: critical pressure. NB 1: When the temperature and the pressure are higher than both of the critical temperature and the critical pressure, the interface between the liquid and the vapor disappears, resulting in a supercritical fluid indistinguishable the liquid and the vapor. NB 2: According to the study of Keiko Nishikawa *et al.* [8], it was found that the boundary line of density fluctuation (dotted line) CD continues on the extension of OC, and the region of the supercritical fluid can be divided into two areas.

Hereupon, by using this P–T diagram we consider a case where the pressure is constant. A horizontal line H is drawn at P = 1 atm in the P–T diagram. This straight line H is parallel to the horizontal axis T. A person, Mr. Honda for example in this case, stands on this straight line H at a low temperature. Mr. Honda is Solid Honda because he is in the solid phase at the low temperature. On this line H, Mr. Honda walks to the right from the low temperature. In other words, Solid Honda is heated under one atmospheric pressure (P = 1). So, he encounters a solid–liquid boundary curve OB. Here (at the mp), Solid Honda transforms into a liquid phase, so that he transforms into Liquid Honda. When he further walks on this line to the right, Liquid Honda encounters the liquid-vapor boundary of curve OC. Here (at the bp), Liquid Honda boils to transform into vapor, and finally disappears. He is invisible over this temperature (bp), but he is around us as Vapor Honda. As described above, when a horizontal line H at a constant pressure is drawn in the P–T diagram and Mr. Honda walks on the straight line H, all the phase transitions can be easily and concretely imagined with a Transformer Honda.

2.3.2.2   *Relationship between polymorphisms in P–T diagram: Enantiotropic relationship and monotropic relationship*

Figures 2.14[A] and 2.14[B] depict schematic P–T diagrams in the cases where crystal polymorphs I and II have an enantiotropic relationship and a monotropic relationship, respectively. In this figure, a solid curve is a boundary of a real phase transition, and a dotted curve is a boundary of a virtual phase transition.

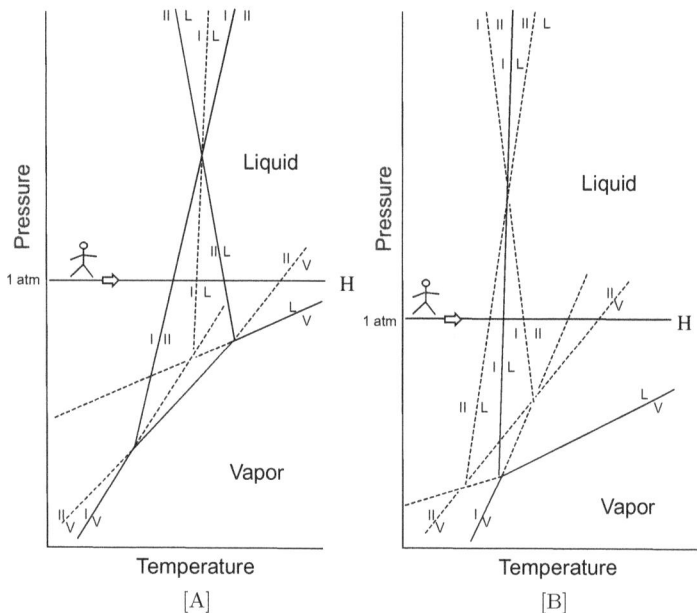

Figure 2.14.   P–T diagrams for [A] enantiotropic relation and [B] monotropic relation.

In Fig. 2.14[A], we will consider the case heating from low temperature when the pressure is constant at 1 atm. You enter into this figure and stand on a straight line H at 1 atm. At low temperatures, you are crystal polymorph I. As you slowly walk to the right on this line H, you reach the realistic border of I–II. At that temperature, you undergo a solid–solid transition and transformed into crystal polymorph II. If you go further to the right, you will reach the realistic border of II–L (liquid). At the melting point of this crystal polymorph II, you melt into liquid L. On the other hand, when you rapidly run to the right on this line H, even if you reach the realistic borderline of I–II, you can not completely transform into crystal polymorph II. On further running to the right, you reach a virtual boundary line of I–L, and here, the crystal polymorph I which remains without being completely transformed melts and becomes liquid L. In this temperature range, the unstable liquid L relaxes and resolidifies into more stable crystal polymorph II. When you furthermore run to the right on this line H, you reach the realistic border of II–L(liquid). At the melting point of this crystal polymorph II, you melt again into liquid L. Therefore, when you run rapidly to the right on this line H, i.e., on rapid heating, **double melting behavior** can be observed. Such phase transition sequence is the same as seen in Figs. 2.10[A] and 2.10[C]. Thus, even using the P–T diagram in this way, it is possible to reproduce the double melting behavior in the case where crystal polymorphs I and II are in an enantiotropic relationship.

Next, in Fig. 2.14[B], we will consider the case heating from low temperature when the pressure is constant at 1 atm. You enter into this figure and stand on a straight line H at 1 atm. At low temperatures, you are crystal polymorph I. As you walk to the right on this line H, you reach the realistic border of I–L. At the melting point of this crystal polymorph I, you melt into liquid L. On the other hand, if you are crystal polymorph II from the beginning at low temperature and you slowly walking to the right on this line H, you reach the virtual boundary of II–L. At this melting point of II, you completely melt to the liquid L. However, in this temperature range, the unstable liquid L slowly relaxes and resolidifies into more stable crystal polymorph I. When you furthermore walk to the right on this line H, you reach the realistic border of I–L. At the melting point of this crystal polymorph I, you melt again into liquid L. Therefore, if you walk slowly on this line H and move to the right, that is, on slowly heating, **double melting behavior** can be observed. Such phase transition sequence is the same as seen in Figs. 2.10[B] and 2.10[D]. Thus, even using the P–T diagram in this way, it is possible to reproduce the double melting behavior when the crystal polymorphs I and II have a monotropic relationship.

As described above, you can reproduce the double melting behavior of crystal polymorphs I and II which have an enantiotropic relationship and a monotropic relationship not only in the G–T diagram but also in the P–T diagram.

However, as can be seen from the P–T diagrams of Figs. 2.14[A] and 2.14[B], in the area where the pressure is much higher, the enantiotropic relationship and the monotropic relationship of crystal polymorphs I and II are reversed. As such, the use of these terms require caution.

2.3.2.3 *Example of G–T diagram*

Figure 2.15 depicts the G–T diagram of HMX (cyclotetramethyl-tetranitramine) having four crystal polymorphs. HMX is widely used as a military and civilian explosive. It is very interesting that the spacecraft "Hayabusa 2" dug an artificial crater on the asteroid Ryugu using the HMX for the first time in human history in April 2019 [9]. In this G–T diagram of HMX, the temperatures at the intersection are the measured values, but the actual slopes are unknown. Generally, when a compound has plural crystal polymorphs like this compound, it shows very complex phase transition behavior. Therefore, it is necessary to draw a G–T diagram to verify whether the phase transition sequence is correct or not. When complex phase transition sequences should be established, we should use a G–T diagram in any cases.

There have been many articles that describe only the enantio-tropic relationship in the phase transition sequence, and that

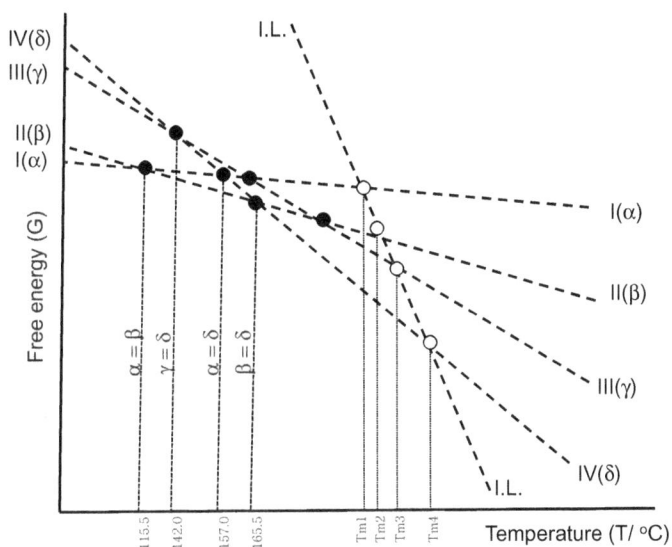

Figure 2.15. G–T diagram for HMX(cyclotetramethylne tetranitramine). The intersection temperatures are measured points, but the actual slopes are unknown. HMX is widely used as an explosive in military and civil applications.

completely ignore the monotropic relationship which causes the multiple melting behavior and relaxation phenomena of metastable polymorphs. In these papers, in differential scanning calorimetry, the first heating measurement using a freshly prepared virgin sample is completely disregarded, and only the second heating and subsequent measurements are adopted. As mentioned above, we have valuable crystal polymorphs that appear only in virgin samples, which give the important measurement results. These first measured results and the complex phase transition behavior can be rationally and surely explained by drawing a G–T diagram. Therefore, all the researchers are required not to discard the first measured results but to utilise them and establish all the phase transitions by drawing a G–T diagram.

2.3.3 *Nomenclature of polymorphs [1]*

International rules have not yet been established for nomenclature of polymorphs. Therefore, according to customs, polymorphs are often named in Roman numerals (I, II, III, IV ...) in the order of stabilities. Polymorph I is the most stable at room temperature. Therefore, as can be seen from the example in Fig. 2.15, the polymorphs are named in Roman numerals I, II, III and IV according to their stabilities at room temperature.

As can be clearly seen from this figure, the polymorphic Roman numerals I, II, III and IV also correspond to their melting points: from the lower one to the higher one. Accordingly, the crystal polymorphs of many compounds are customarily named as Roman numerals I, II, III, IV ..., in the order of their melting points. However, if in further studies we find a new polymorph between polymorphs II and III, it will have to name it as 2.5. Therefore, the nomenclature of using Roman numerals is not a strict rule. Some researchers in fact prefer to name polymorphs using Greek letters α, β, γ, δ,....

Confusingly, many liquid crystal researchers name liquid crystal polymorphs in the reverse order from the highest melting point to the lowest one, in Arabic numerals (1, 2, 3, 4, ...) Thus, there is no unified rule in the naming system in science. Therefore, younger generation of researchers should discuss with IUPAC and related organizations to establish uniform naming rules in all fields in science. For the time being, it is necessary to read each of the paper to confirm what rules are used by the authors.

2.3.4 *Composition diagram*

The composition diagrams include a temperature-composition phase diagram (T–X diagram) and a pressure-composition phase diagram (P–X diagram). Here, we will consider the phase transition using a temperature-composition phase diagram (T–X diagram).

2.3.4.1 *In the case of vapor–liquid phase diagram*

*An ideal case when components A and B do not dissolve each other at all

For this case, the temperature-composition phase diagram (T–X diagram) is as shown in Fig. 2.16. The liquid separates into two phases L_A and L_B at any composition. Curves T_{Ab}–E and T_{Bb}–E in the figure are called condensation curves, and point E is called an azeotropic point.

*In a case when components A and B partially dissolve for each other

The above-mentioned T–X diagram in Fig. 2.16 is a very ideal case when components A and B do not dissolve each other at all. However, usually, components A and B partially dissolve each other. Each component does not dissolve so much at low temperatures, but the solubility increases with increasing temperature; over a certain temperature Tc, they become mutually soluble in all compositions. Therefore, the mutual solubility curve becomes like as Fig. 2.17. The point Tc in this figure is called upper critical solution temperature. As can be seen from this figure, under the mutual solubility curve, there are two phases of L_A and L_B, while over the curve, they are uniformly dissolved and become one phase which is called a solution.

When Tc of the mutual solubility curve is considerably higher than the azeotropic point T_E, as shown in Fig. 2.18, the solution exists only in two limited regions near the pure component A or the pure component B.

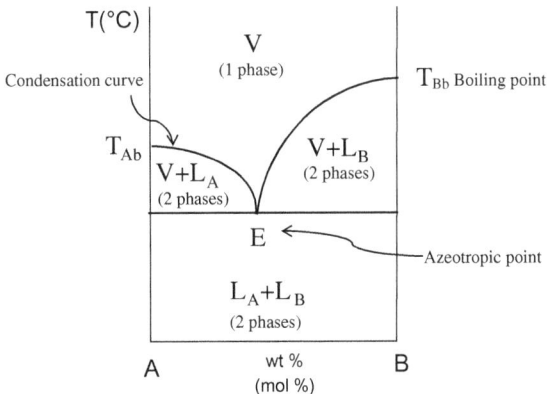

Figure 2.16. Temperature-composition phase diagram (T–X diagram) in an ideal case when components A and B do not dissolve at all. V = vapor and L = liquid.

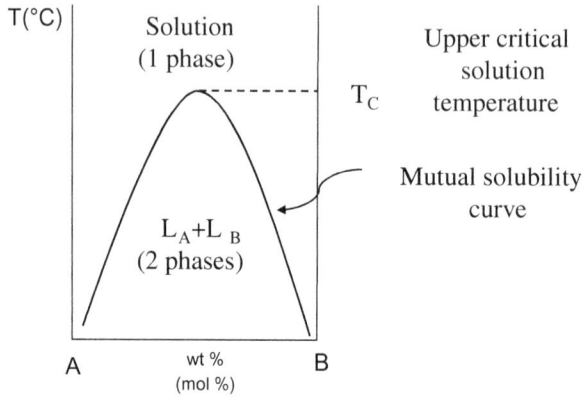

Figure 2.17. In a case when two components A and B are partially soluble with showing an upper critical solution temperature, T_C.

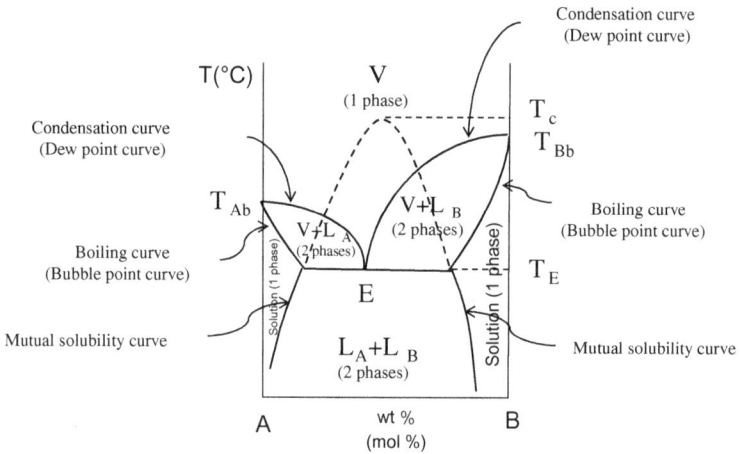

Figure 2.18. Vapor-liquid composition diagram. In a case when components A and B partially dissolve. V = vapor and L = liquid.

*Various vapor–liquid composition phase diagrams depending on the relative positional relationship between Tc and T_E**

You can easily understand that various vapor–liquid composition phase diagrams can be obtained by depending on the relative positional relationship between Tc and T_E, as illustrated in Figs. 2.19(A)~(F).

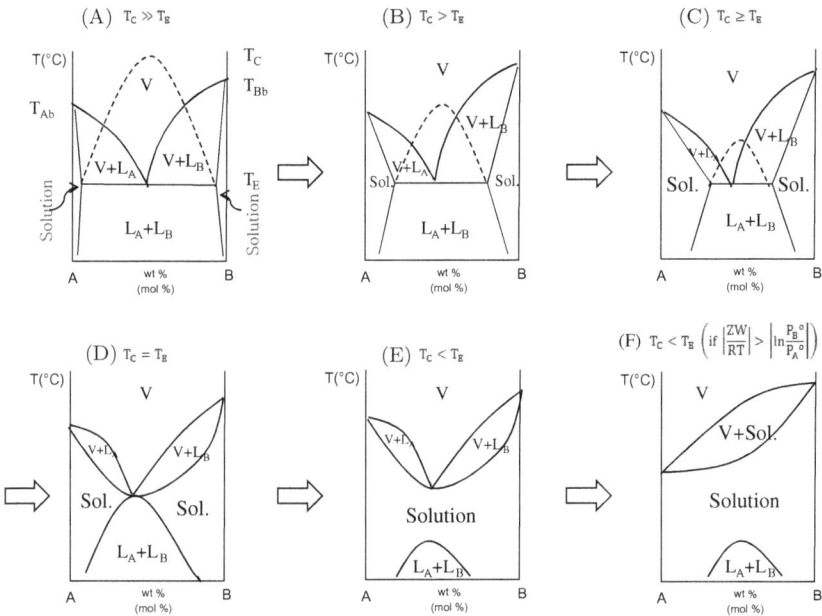

Figure 2.19. Depending on the relative relationship between T_c (upper critical solution temperature) and T_E (Azeotropic point), various vapor phase-liquid phase component (T–X) phase diagrams can be obtained.

As shown in Fig. 2.19(F), no azeotropic mixture is given when $\left|\frac{Zw}{RT}\right| > \left|\ln\frac{P_B^0}{P_A^0}\right|$. This explanation is beyond the scope of this book, so avid readers should look at an advanced book, for example, *The Nature of Solution I*, by Fujishiro and Kuroiwa, pp. 124–127 [10].

2.3.4.2 *Solid–liquid composition phase diagram*

Thermodynamically, it can be considered exactly as in the case of vapor–liquid. The composition phase diagram is as shown in Fig. 2.20. As can be seen from this figure, the solid phase under the eutectic point temperature T_E and under the mutual solubility curve is separated into two phases of K_A and K_B. However, in two limited regions near pure component A or pure component B, over the mutual solubility curves, there is a solid phase in which each component is uniformly dissolved to show one phase. This is called **solid solution**.

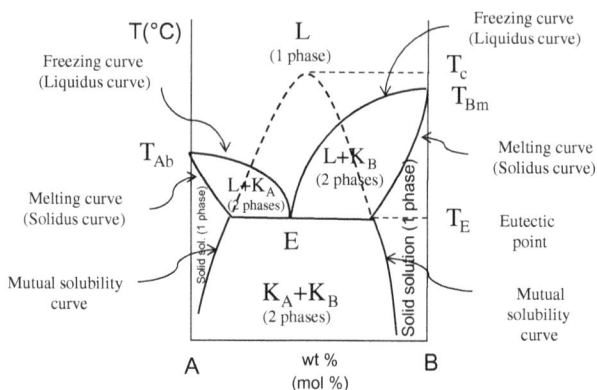

Figure 2.20.   Solid–liquid composition diagram. In a case when components A and B partially dissolve. L = liquid and K = solid.

Therefore, a solid phase $(K_A + K_B)$ in which two phases are separated is not called as a solid solution. In the solid phase $(K_A + K_B)$, the particles of A and B are dispersed in the solid, so that this state is not a solid solution but a **colloid**. The **solution (solution, solid solution)** always means mixing and dissolving occur at the molecular (atomic) level. Note that these terms are often misused.

2.3.4.3   *Phase rule (For details, see, e.g., Atkins Physical Chemistry 8th ed., Chapter 8)*

When setting parameters as follows, F: the degree of freedom of the system (the number of intensive variables), C: the number of components, P: the number of phases,

$$F = C - P + 2 \qquad (2.8)$$

which is called as phase rule. This rule holds universally without exception. Since intensive variables determine the state of the system, three parameters of temperature, pressure, and component composition are enough for the consideration. We look at the usefulness of the phase rule in the following examples.

Example 1. Boiling point and melting point are fixed points under atmospheric pressure.

Proof. As shown in Fig. 2.16, at the boiling point T_{Ab} of component A, $C = 1$ and $P = 2$, and the pressure is specified as atmospheric pressure, so $F = 0$ and the temperature does not change.

Thus, if the phase rule is used, it can be easily proved that the boiling point is a fixed point under atmospheric pressure.

Example 2. Under atmospheric pressure, the azeotropic point and the eutectic point are fixed points.

Proof. $C = 2$ and $P = 3$ at point E in Fig. 2.16, and the pressure is specified as atmospheric pressure, so $F = 0$. Therefore, at the eutectic point, the temperature and composition are both constant and fixed.

Example 3. The triple point is a fixed point.

Proof. In Fig. 2.13, $C = 1$ and $P = 3$ at point O, so $F = 0$. Therefore, at the triple point the temperature and pressure are both constant and fixed.

NB: There is no exception because phase rule is universal. Several tens of nm to one hundred nm of a **thermal equilibrium periodic structure** is formed by copolymer, water, oil, mixed surfactant system, etc., and it has been called a **microphase separation structure** since around 2000 [11]. However, this term claims that microphase separation occurs in one phase, so that it clearly violates phase rule.

Depending on the temperature and composition ratio, the **thermal equilibrium periodic structure** of copolymer, water, oil and mixed surfactant systems shows lamellar, hexagonal cylinder, body-centered cubic (BCC), and gyroid structures. These structures are completely equivalent to the pure liquid crystalline phases of smectic (lamella), hexagonal columnar, cubic, gyroid structures. These liquid crystal phases are well and strictly defined in the research field of liquid crystals. Therefore, the **thermal equilibrium periodic structures** which are claimed by the researchers in the copolymer and surfactant fields are exactly the same as the liquid crystal phase structures in the liquid crystal research field.

Each **periodic structure** is in thermal equilibrium between other structures. In other words, phase transition can occur between these phases. Therefore, the **thermal equilibrium periodic structure** must coincide with a single phase as the same as the liquid crystal phase, so that it should not be separated into two phases in a phase, strictly judging from the viewpoint of phase rule. Nevertheless, this term claims that **microphase separation** would occur in one phase, which obviously violates universal phase rule.

There is no exception for phase rule. Hence, the term of **microphase separation** should be corrected to **microlayer separation**, consistently with universal phase rule.

2.3.4.4 *Le Chatelier–Schröder equation (For details, see, for example, Physical Chemistry by Moore, Chapter 7 [12])*

The freezing point depression curve and the eutectic point can be theoretically calculated using Le Chatelier–Schröder equation in the temperature–composition phase diagram of the binary system (for example, the T–X diagram in Fig. 2.20).

In the two-component system, for component 1, setting as X_1: mole fraction, ΔH_{f1}: enthalpy of melting, T_1: freezing point (= good as melting point), R: gas constant,

$$\ln X_1 = \frac{\Delta H_{f1}}{R}\left(\frac{1}{T_1} - \frac{1}{T}\right) \tag{2.9}$$

For component 2, setting as $X_2 (= 1 - X_1)$: mole fraction, ΔH_{f2}: enthalpy of melting, T_2: freezing point,

$$\ln(1 - X_1) = \frac{\Delta H_{f2}}{R}\left(\frac{1}{T_2} - \frac{1}{T}\right) \tag{2.10}$$

Equations of 2.9 and 2.10 are called Le Chatelier–Schröder equations.

The freezing point depression curve and the theoretical values of the eutectic point can be determined as follows. First, the enthalpy ΔH_{f1} and ΔH_{f2} of melting of the pure components 1 and 2 and the melting points T_1 and T_2 are measured with a differential scanning calorimeter. Next, the temperatures T are stepwise calculated by using the above two Equations 2.9 and 2.10, when X_1 is substituted into these equations for each of the values increased by every 0.1 for $X_1 = 0 \sim 1.0$. The obtained (X_1T) values are plotted in a T–X diagram to draw two theoretical freezing point depression curves. Since the intersection of the two curves is the eutectic point, you can read the temperature and composition from the intersection. You cannot solve Equations 2.9 and 2.10 algebraically, similarly to simultaneous equations, so it is better to find the intersection graphically as described above (see Question 13 at the end of this chapter).

2.3.4.5 *Binary phase diagram with polymorphism (Reason 1 why many eutectic points are observed)*

Figure 2.21 shows the composition phase diagrams of δ-hexachlorocyclohexane having two crystal polymorphs and α-hexachlorocyclohexane having only one crystal form. As can be seen from this figure,

$$\text{Number of eutectic points} = 2 \times 1 = 2$$

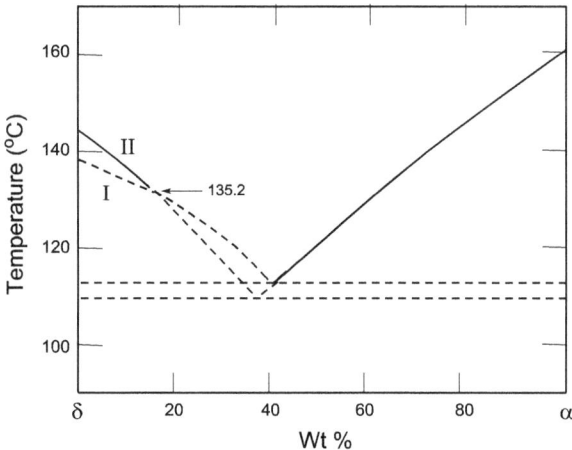

Figure 2.21. Composition diagram for the δ- and α-hexachlorocyclohexanes. The δ-isomer has two crystalline polymorphs.

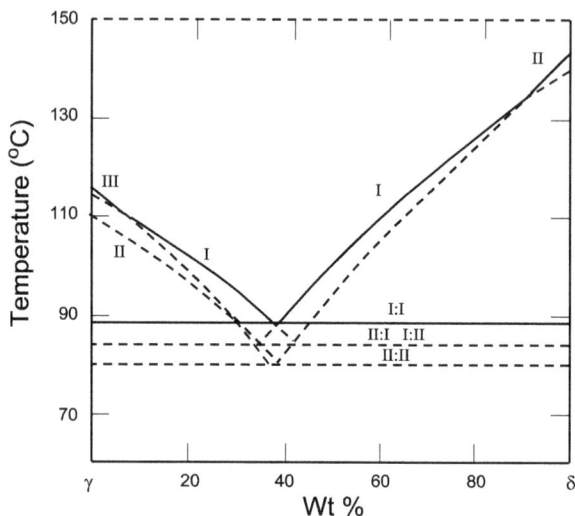

Figure 2.22. Composition diagram for the γ- and δ-hexachlorocyclohexanes. The γ- and δ-hexachlorocyclohexanes have three and two crystalline polymorphs, respectively.

Figure 2.22 shows the composition phase diagrams of γ-hexachlorocyclohexane having three crystal polymorphs and δ-hexachlorocyclohexane having two crystal polymorphs. As can be seen from this figure,

Number of eutectic points $= 3 \times 2 = 6$

(Theoretically, six eutectic points should be observed.)

From the above examples, in the two-component phase diagram, the number of eutectic points is theoretically observed as many as the number of polymorph I multiplied by the number of polymorph II.

2.3.4.6   *Two-component phase diagram when forming an adduct (intermolecular compound) (Reason 2 why many eutectic points are observed)*

*When forming an [AB] type of intermolecular compound**

Figure 2.23 is a binary phase diagram of electron acceptor 1, 3, 5-trinitrobenzene and electron donor phenanthrene. Such an

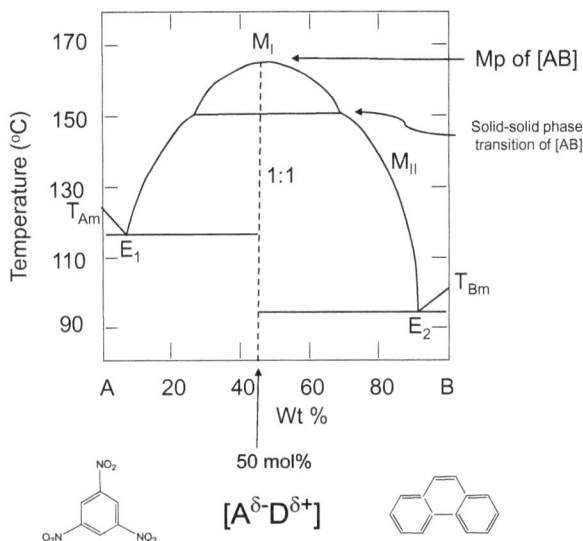

Figure 2.23. Composition diagram for 1,3,5-trinitrobenzene (A) and phenan-threne (B) showing polymorphism of the addition compound.

electron acceptor [A] and an electron donor [D] often form an $[A^{\delta-}D^{\delta+}]$ charge transfer complex type of intermolecular compound. Therefore, when the concentration of each of A and D is 50 mol%, a (1:1) intermolecular compound is formed. The melting point of this intermolecular compound newly appears. Accordingly, two different composition phase diagrams exist side by side in this figure, so that two eutectic points E_1 and E_2 can be observed.

*When forming an [AB₂] type of intermolecular compound

In this case, the melting point of the intermolecular compound is observed at 33.3 mol% as shown in Fig. 2.24.

As can be seen from these two examples mentioned above, the ratio of intermolecular compound can be determined by drawing a two-component phase diagram. This is one of the applications of two-component phase diagrams. In the 1980's, many charge transfer complexes were actively synthesized in order to obtain the organic superconductors, and at that time such two-component phase diagrams were often drawn to determine the component ratios.

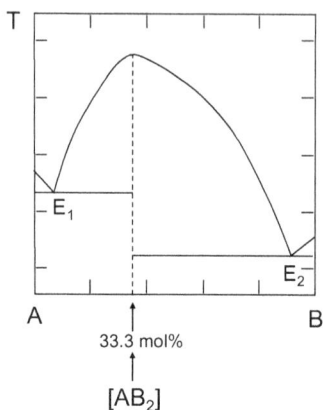

Figure 2.24.   In the case of forming an intermolecular addition compound of [AB$_2$] type, the ratio can be determined from the component diagram (application of the T–X phase diagram).

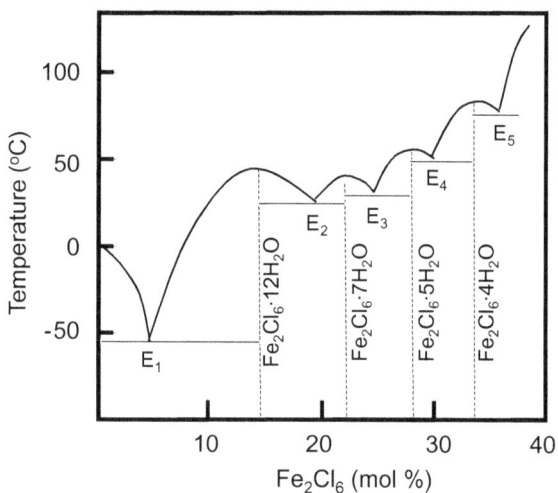

Figure 2.25.   Composition diagram for Fe$_2$Cl$_6$-H$_2$O.

*When forming many kinds of intermolecular compounds

When Fe$_2$Cl$_6$ is recrystallized from water, four kinds of inter-molecular compounds Fe$_2$Cl$_6 \cdot n$H$_2$O ($n = 12, 7, 5, 4$) incorporating crystal water are obtained as shown in Fig. 2.25. As can be seen from this figure, five eutectic points E$_1 \sim$ E$_5$ are observed.

NB: Na_2SO_4 is commonly used in chemical laboratories as a desiccant

Anhydrous Na_2SO_4 can incorporate at most 10 water molecules as crystal solvent to precipitate out as an intermolecular compound $Na_2SO_4 \cdot 10H_2O$. The anhydrous Na_2SO_4 is often used as a desiccant because it can incorporate such a large number of water molecules and still exists as crystals.

2.4 Solution Phase Transformation

2.4.1 *Measuring apparatus*

The solution phase transition is observed with an apparatus [1, 2] like Fig. 2.26.

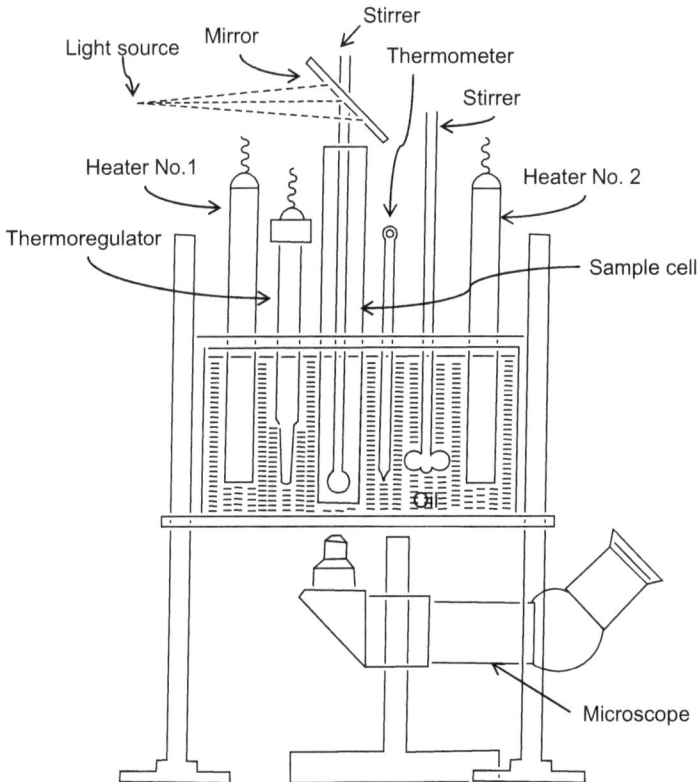

Figure 2.26. Microscope setup for observing solution phase transformations [1, 2].

2.4.2 *Measurement principle*

As shown in Fig. 2.27, two crystal polymorphs I and II are directly adjacent to each other. When it is heated or cooled, the solid–solid phase transition temperature of polymorphic I ⇆ polymorph II can be examined whether the boundary proceed to the right or the left.

However, superheating and supercooling of the solid–solid phase transition often take place in solid phases, so that it is very difficult to find the accurate phase transition temperature. This is because molecular rearrangement is less likely to occur in the solid phase.

Accordingly, as shown in Fig. 2.28, both crystal polymorphs I and II are placed in the bottom of flat test tube with a small amount of poor solvent. Since only a small amount of poor solvent is added,

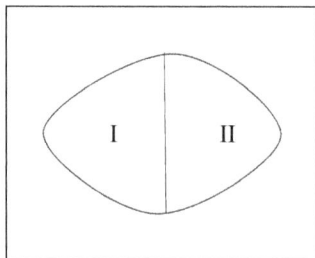

Figure 2.27. Schematic representation of a solid–solid phase polymorphic transformation. We can observe the transformation temperature by the change of the boundary line between solid polymorphs I and II.

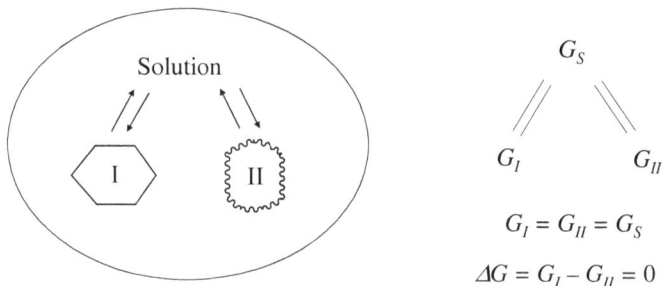

$$G_I = G_{II} = G_S$$

$$\Delta G = G_I - G_{II} = 0$$

Figure 2.28. Thermodynamic principle of solution phase transformation.

a large amounts of the crystal polymorphs I and II can co-exist in the solution without completely dissolving.

When such a sample is observed with heating or cooling by using a special microscope like Fig. 2.27, the phase transition temperature of polymorph I $\leftrightarrows$ polymorph II can be very accurately determined by judging whether crystal polymorph I or II becomes fat or thin.

The thermodynamic principle is proved as follows.

First of all we will consider the case when both crystal polymorph I and crystal polymorph II become neither fat nor thin.

At this time, since the crystal polymorph I and the solution are in equilibrium, their chemical potentials are equal.

$$G_I = G_s$$

Also, since the crystal polymorph II and the solution are in equilibrium, their chemical potentials are equal.

$$G_{II} = G_s$$

$$\therefore G_I = G_{II} = G_s$$

$$\therefore \Delta G = G_I - G_{II} = 0$$

$\Delta G = 0$ occurs at the solid–solid transition temperature of crystal polymorph I $\rightleftarrows$ crystal polymorph II.

From such a thermodynamic principle, the phase transition of crystal polymorph I $\rightleftarrows$ crystal polymorph II can also take place via the solution. This is called as **solution phase transformation**. The solid–solid phase transition temperature can be determined extremely accurately by using solution phase transformation. It is because the rearrangement of molecules can easily take place via the dilute solution.

2.4.3 *Solubility curve*

Figure 2.29 illustrates a G–T diagram of two crystal polymorphs I and II showing a solid–solid phase transition at 115.5°C. For this case we will consider the stabilities and solubilities of these

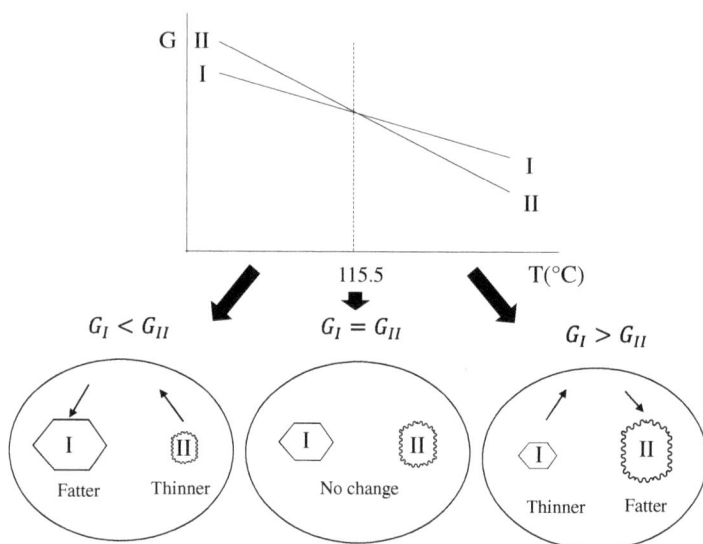

Figure 2.29. An example of solubility curves of crystalline polymorphs I and II. In this case a solid–solid transition occurs between polymorphs I and II at 115.5°C. Below 115.5°C, polymorph I is more stable than polymorph II, because $G_I < G_{II}$ in the G–T diagram. Above 115.5°C, polymorph II is more stable than polymorph I, because $G_I > G_{II}$ in the G–T diagram. At 115.5°C, polymorphs I and II are in a phase equilibrium, because $G_I = G_{II}$. Therefore, in a poor solvent below 115.5°C, polymorph I becomes fatter but polymorph II becomes thinner in the solution. Above 115.5°C, polymorph I becomes thinner but polymorph II becomes fatter. At 115.5°C, neither polymorph I nor polymorph II change. Accordingly, solubilities of the crystalline polymorphs in a solution depends on the stabilities in the corresponding G–T diagram.

polymorphs. As can be seen from this G–T diagram, below 115.5°C crystal polymorph I is a stable polymorph, because $G_I < G_{II}$. On the other hand, over 115.5°C crystal polymorph II is a stable polymorph, because $G_I > G_{II}$. At 115.5°C, crystal polymorph I and crystal polymorph II are in the phase equilibrium, because $G_I = G_{II}$.

With these things in mind, the appearance of the solution phase transition is schematically illustrated below the G–T diagram.

Below 115.5°C crystal polymorph I is a more stable polymorph, so that crystal polymorph I becomes fat and crystal polymorph II becomes thin in the solution. Over 115.5°C crystal polymorph II is a more stable polymorph, so that crystal polymorph I becomes thin

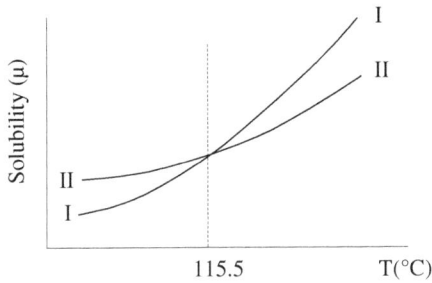

Figure 2.30. Schematic solubility curves for two polymorphic forms, I and II.

and crystal polymorph II becomes fat in solution. At 115.5°C, crystal polymorph I and crystal polymorph II are in phase equilibrium, so both polymorphs become neither fat nor thin.

Therefore, it can be seen that below 115.5°C the solubility of crystal polymorph II is greater while over 115.5°C that of crystal polymorph I is greater.

Accordingly, the corresponding solubility curves of crystal polymorphs I and II can be illustrated as Fig. 2.30. This solubility curves in Fig. 2.30 corresponds well to the G–T diagram in Fig. 2.29. As can be seen from these figures, the lines in G–T diagram are always downward, whereas the solubility curves oppositely become always upward. It is because solubility becomes better with increasing temperature. The solubility curves can be easily predicted from the G–T diagram. Moreover, it can easily predict that the more unstable the polymorph becomes at room temperature, the higher the solubility becomes.

Therefore, it can be easily deduced from the GT diagram of HMX in Fig. 2.15 that at room temperature polymorph $IV(\delta)$ shows the highest solubility and polymorph $I(\alpha)$ shows the lowest solubility, and that the solubility becomes higher in an order of $I(\alpha) \rightarrow II(\beta) \rightarrow III(\gamma) \rightarrow IV(\delta)$. Even for the same drug, the polymorph having the highest solubility at room temperature shows the highest medicine efficacy. It is because that the polymorph having the highest solubility can maximize the blood concentration. Therefore, crystal polymorphism is a major pharmaceutical problem.

2.5 How to Prepare Metastable Polymorph [1]

1. Place the compound in the temperature range where the desired polymorph is the stable form. It can be prepared quickly by using solution phase transition. cf. Fig. 2.15.
2. Quench the melt (IL). By supercooling, a metastable polymorph can be prepared before a stable polymorph appears.

2.5.1 *Example A prepared from solution*

HMX (I): a 200 ml of hot saturated acetone solution is spontaneously cooled to room temperature. (Very slow cooling, low supersaturation)

HMX (II): a 50 ml of the same hot solution is cooled to room temperature with gentle stirring.

HMX (III): a 30 ml of the same hot solution is cooled in a cold water bath with stirring.

HMX (IV): a 5 ml of solution in a test tube is placed in a dry ice-ethanol bath or poured onto crushed ice.

In all the cases, the crystals should be filtered immediately to avoid taking place of solution phase transition to stable HMX (I) above 115.5°C.

From the above operations, we can get the following knack of preparing metastable polymorphs.

> In order to obtain a more unstable polymorph, it is necessary to quench the hotter solution more quick.

2.5.2 *Stability of metastable polymorphs*

The decisive factor is crystal size.

> Metastable polymorphs are more unstable in larger crystals while more stable in smaller crystals.

(Example) HMX (IV)

- Crystals having more than 1 mm in size transform into HMX (I) within a few hours.
- Small crystals having less than $50\,\mu$m in length and $10\,\mu$m in thickness remain as they are, without transforming for 20 years.

This reason can be explained as shown in Fig. 2.31.

This figure schematically illustrates a square plate-shaped metastable crystal polymorph IV having 10 mm in one side. It is assumed that spontaneous transformation of this metastable polymorph IV to the stable polymorph I will take place at two points per $1\,\text{cm}^2$. Once the spontaneous transformation takes place at these two points, the transformation gradually spreads from these points, and the entire crystal transforms within several hours. However, when the crystal having a side length of 10 mm is divided into four as shown in the figure, two out of four parts do not undergo spontaneous transformation. When the crystal is further divided into 16 parts, 14 out of the 16 parts do not undergo spontaneous transformation. Therefore, when this metastable polymorph IV is prepared a very fine crystal powder having $10\,\mu$m in one side, only two out of one million

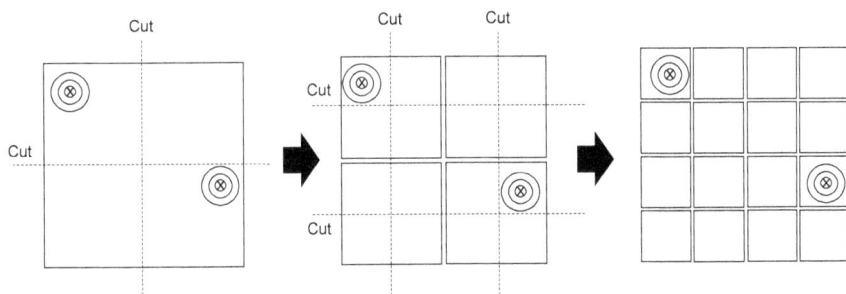

Figure 2.31. Stability of metastable crystalline polymorph. X shows a spontaneous transformation point from a metastable polymorph to a more stable polymorph. Since the spontaneous transformation generally occurs in a certain probability per volume, the smaller the crystalline size becomes, the bigger the survival possibility of metastable crystalline polymorph becomes.

parts undergo spontaneous transformation. Since this transformation is negligible, no spontaneous transformation will actually take place.

Thus, the spontaneous transformation generally occurs only with a certain probability per volume, so the smaller the crystal size becomes, the dramatically more stable the metastable crystal polymorph becomes. Therefore, highly drug efficacy of metastable polymorphs must be stored as very fine crystalline powders. If it is stored as large crystals, its drug efficacy will rapidly degrade by changing to a less stable polymorphic form. Pharmaceutically, this fine powdering is a very important storage technology.

2.5.3 *Example B prepared from melt*

DINA (nitramine dioxyethylnitramine dinitrate) has four crystal polymorphs with different melting points as follows.

DINA (I): 52°C. higher by +0.005°C than that of DINA (II)
DINA (II): 52°C.
DINA (III): 38°C.
DINA (IV): 30°C

Each of these crystal polymorphs can be obtained, when the supercooled melt is strongly pressed from on the cover glass with a tip of a needle at different temperatures to form a nucleus (seed crystal).

If this operation is carried out at the following different temperature, the desired crystal polymorph can be obtained.

At 45 to 40°C, DINA (II) is obtained.
At 35 to 30°C, DINA (III) is obtained.
Below 30°C, DINA (IV) is obtained.

From the above operations, we can get the following knack of preparing metastable polymorphs.

> From the more supercooled liquid, the more unstable polymorph can be obtained.

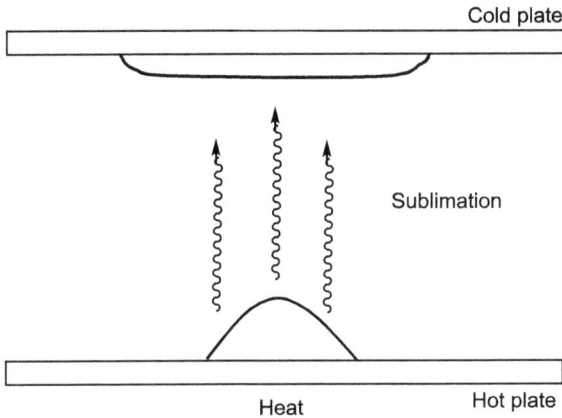

Figure 2.32. Formation of a crystalline polymorph from the vapor by sublimation. The colder the upper plate becomes, the more unstable polymorph can be obtained.

2.5.4 *Example C prepared from vapor*

Figure 2.32 shows a schematic diagram of sublimation purification. The lower the temperature of the cold plate at the top of the figure becomes, the more unstable polymorph can be obtained. This also indicates that it is necessary to quench the hot vapor more severely in order to obtain a more unstable polymorph.

2.6 Verification Method Whether Polymorphs Exist or Not

Suppose that you successfully discover a new medicine after many years of your research. You would sincerely hope that this medicine would have high drug efficacy as possible to save many of the afflicted patients. Therefore, you must hope to find the crystal polymorph showing the highest drug efficacy as possible even in the same compound.

There is also an obligation to report all the melting points of the crystal polymorphs in this compound when applying as a new drug to the country. Therefore, it is necessary to have a verification method whether crystal polymorphism exists or not in this new compound. The verification should be as follows.

2.6.1 Dissolve a small amount of the compound completely under crossed nicols of a polarization microscope, and cool it to observe solidification. If the solid–solid transition occurs after spontaneous solidification, the compound has at least two polymorphs.

$$I.L. \xrightarrow{Cooling} K_1 \xrightarrow{Cooling} K_2$$

2.6.2 Heat the compound and observe if solid–solid transition occurs during heating. The opportunity for transformation is higher for larger crystals (cf. 2.5.2).

$$K_1 \xrightarrow{Heating} K_2$$

2.6.3 Sublimate a small amount of the compound and observe whether a solution phase transition occurs between the sublimate and the original sample.

2.6.3.1 If two are polymorphic, the metastable form dissolves better and the stable form grows. Growth of the stable form continues until the metastable form disappears (cf. 2.4).

2.6.3.2 If two compounds are not polymorphic but different compounds, one dissolves but another does not grow.

2.6.3.3 If two are exactly the same polymorphs, no change occurs.

2.6.4 Place the excess solid compound in a small amount of solvent kept at a temperature as close as possible to the melting point of the compound, and keep it for hours and isolate quickly. It is likely that the original metastable polymorph has completely changed to the high temperature polymorph. The product and the original compound are verified in method (2.6.3).

2.6.5 Recrystallize the compound by quenching the small amount of solution very quickly. There is a high possibility to produce a metastable polymorph (cf. 2.5.1). The product and the original compound are verified in method (2.6.3).

2.7 Confirmation Method Whether Two Given Samples are Polymorphs of the Same Compound

Suppose that you are a forensic examiner in a national forensic science and research institute and you have to identify whether the compound, such as a drug, poison, or explosives left behind on the crime scene, are the same compound found in a hideout of the criminals. First, you will identify the molecular structures by using UV-visible spectrometer, infrared spectrometer, proton nuclear magnetic resonance spectrometer, etc. in order to prove that two compounds in the criminal scene and the hideout have the identical molecular structure. However, even if you can prove in such a molecular level, it may be still necessary to further prove that both are in the same molecular assembly state. If it is proven that the same polymorph of the same compound was used, it can be very solid evidence.

If you are required to confirm whether such two given samples are the same polymorph of the same compound, the following method can be used.

2.7.1 If two samples have different crystal systems, i.e., axial ratios, refractive indices, densities, and X-ray powder patterns, and if they show solid–solid phase transition or solution phase transition between them, these two crystals are the polymorphs of the same compound.

2.7.2 When two kinds of mixed crystals (A + B) are heated up to the first lower melting point, the resulting liquid is kept at the temperature; the melt completely recrystallizes into another crystal by seeding the remained crystals (B), and on further heating, they all completely melt at the higher melting point of B. In this case, two kinds of the crystals are the polymorphs of the same compound. The state changes become as illustrated in Fig. 2.33.

2.7.3 If the melts of the two samples show exactly the same properties (refractive index, temperature coefficient of refractive index) and the two crystallinity differ as in (2.7.1), The two crystals are polymorphs of the same compound (cf. 2.1).

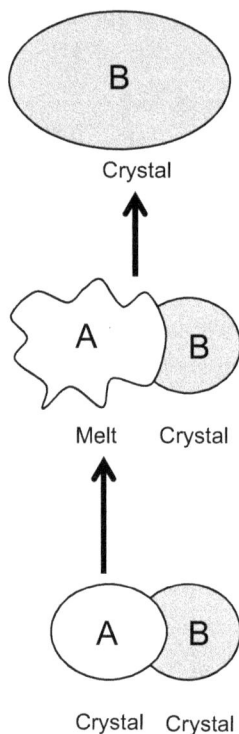

Figure 2.33. When two kinds of mixed crystals (A + B) are heated up to the first lower melting point, the resulting liquid is kept at the temperature; the melt completely recrystallizes into another crystal by seeding the remained crystals (B), and on further heating, they all completely melt at the higher melting point of B. In this case, two kinds of the crystals are the polymorphs of the same compound.

2.7.4 Carry out a series of operations as shown in Fig. 2.34. One of the two crystalline samples A and B is heated to melt. The melt is subcooled slightly below its melting point. Onto this supercooled liquid, two crystalline samples A and B are inoculated at two different locations. Each of the crystals grows and collides with each other to form a boundary. The boundary is observed with a polarization microscope. If one of the two crystalline forms continues to grow through the

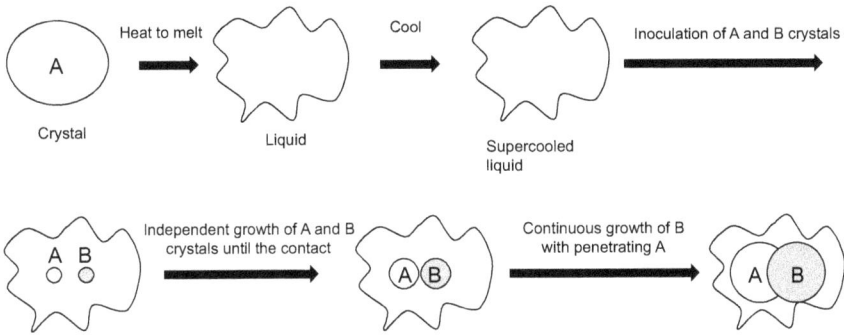

Figure 2.34. Proof of polymorphism. One of the two crystalline samples A and B is heated to melt. The melt is subcooled slightly below its melting point. Onto this supercooled liquid, two crystalline samples A and B are inoculated at two different locations. Each of the crystals grows and collides with each other to form a boundary. The boundary is observed with a polarization microscope. If one of the two crystalline forms continues to grow through the other crystal, these two crystalline samples A and B are the polymorphic forms of the same compound.

other crystal, these two crystalline samples A and B are the polymorphs of the same compound (cf. Fig. 2.27).

2.7.5 Slide the solvent into two samples between the two cover glasses and observe the solution phase transition under a microscope. If a solution phase transition is observed between the two samples, then the two samples are crystal polymorphs of the same compound (cf. 2.6.3).

2.7.6 Two cases are shown in Fig. 2.35.

Case I: When two kinds of crystals are heated together and one crystal transforms by the solid–solid transition and on further heating, they melt at the same temperature. It strongly suggests that these two crystals may be the polymorphs of the same compound. But it is not proof because these two different compounds accidentally have the same melting points. Therefore, there still remains the possibility of two different compounds.

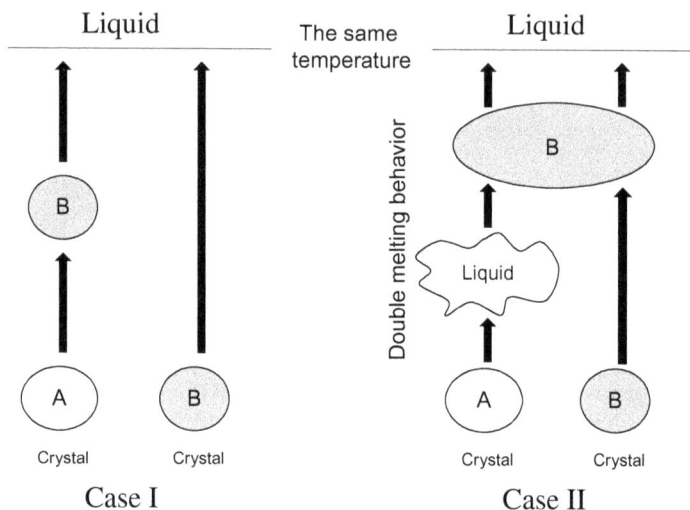

Figure 2.35.   Proof of polymorphism by using double melting behavior. Case I: When two kinds of crystals are heated together and one crystal transforms by the solid–solid transition and on further heating, they melt at the same temperature. It strongly suggests that these two crystals may be the polymorphs of the same compound. But it is not proof because these two different compounds accidentally have the same melting points. Therefore, there still remains the possibility of two different compounds. Case II: When two kinds of crystals are heated together and one crystal form melts, then the melt completely recrystallizes by seeding another crystal form, and on further heating, they all melt at a single melting point. This is a perfect proof that these two crystals are the polymorphs of the same compound.

Case II: When two kinds of crystals are heated together and one crystal form melts, then the melt completely recrystallizes by seeding another crystal form, and on further heating, they all melt at a single melting point. This is a perfect proof that these two crystals are the polymorphs of the same compound.

2.7.7 As shown in Fig. 2.36, if two well-mixed crystalline samples (A + B) are sublimed, it may result to crystallize as one polymorph (C). This is a proof of crystal polymorphism by **"vapor phase transition."**

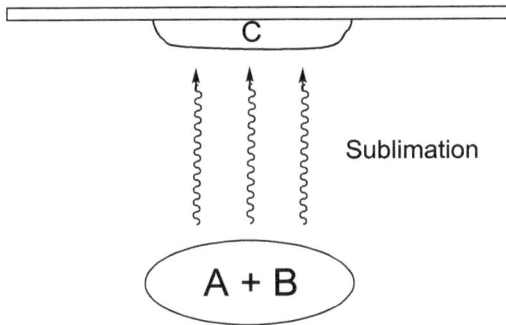

Figure 2.36. Vapor phase transformation. If two well-mixed crystalline samples (A + B) are sublimed, it results to crystallize as one polymorph (C). This is a proof of crystal polymorphism by "vapor phase transition."

2.8 Application of Polymorphism [2]

2.8.1 *Suppository*

A suppository is a solid formed by mixing the drug with theobroma oil and molding it into a rocket shape. If the patient can not take the drug orally in infants or young children, this suppository should be inserted into the anus for ingestion (cf. Fig. 2.37).

The base of suppository is theobroma oil, which is an oil taken from cacao and is usually used as a source of chocolate. Theobroma

Figure 2.37. Shape of a suppository. A rocket-shaped suppository is prepared by mixing a drug with theobroma oil. It is served by injection into the anus, when infants or young children cannot take the drug orally. Ref. https://www. ayumi-pharma.com/medical/commons/drug/product/photo/m_p13.php.

oil has six crystal polymorphs with different melting points. The lowest melting point is $17.3°C$ for polymorph I and the highest melting point is $36.3°C$ for polymorph VI [13]. For the chocolate or suppository we use polymorphic form V having a melting point at $33.8°C$. Chocolates and suppositories easily melt in the mouth and anus of our body, because these temperatures are just around $33–34°C$. If the crystallization is carried out at $27°C$, the product will melt even by touching with a finger. So the crystallization should be carried out at $33°C$ in order to prepare polymorph V. Therefore, suppositories are prepared by mixing medicine such as antipyretic with theobroma oil. Once the mixture is melted at 60 to $70°C$ the melt is poured into a rocket-shaped mold kept at $33°C$ to solidify. When such suppositories are carelessly left in a warm room for a long time, it transforms into the most stable polymorph VI which will not be melted in the mouth or anus. For this reason when you receive a suppository at the pharmacy you are instructed to put it in the refrigerator immediately after your usage. If the suppository has transformed into the most stable polymorph VI, it will not melt in the body and cause an awful situation. Young parents having infants need to be careful about storage of suppositories.

Thus, suppositories utilize crystal polymorphism.

2.8.2 *How to make rapid efficacy insulin and slow efficacy insulin*

The duration of insulin efficacy is controlled by the degree of crystallinity. Insulin is reacted with zinc chloride and the resulted zinc complex is used as the medicine.

- Rapid efficacy insulin: prepared as amorphous form of fine powder.
- Slow efficacy insulin: composed of 70% of crystals and 30% of amorphous form.

Amorphous refers to a glassy liquid or supercooled liquid, which is more unstable than the most unstable crystal polymorphs and exhibits much higher solubility.

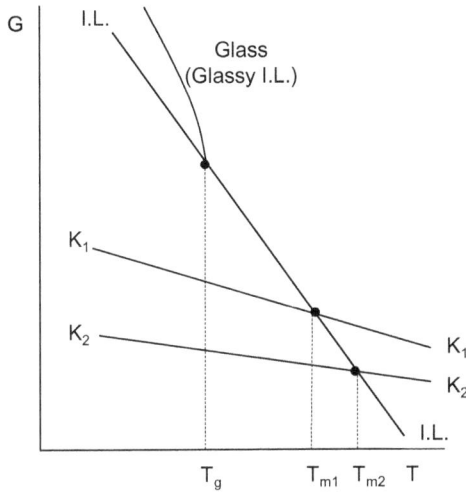

Figure 2.38. A schematic G–T diagram of a glassy liquid (Glassy IL) [14] originated from an isotropic liquid (IL).

Figure 2.38 schematically depicts a G–T diagram of a glassy liquid (Glassy IL) [14] produced from an isotropic liquid (IL). When the IL is rapidly quenched, the supercooled liquid solidifies with remaining the liquid structure below the glass transition temperature (Tg) without crystallization. As can be seen from this G–T diagram, Glassy IL is much more unstable than crystal polymorphs K_1, K_2 and even IL. Therefore, the amorphous insulin is enormously soluble at room temperature, so that the concentration in blood rises instantaneously. Thus it becomes a rapid efficacy insulin. On the other hand, both the crystal polymorphs K_1 and K_2 have much lower solubilities at room temperature, so that they become slow efficacy insulin because they dissolve slowly and gradually in blood.

Patients suffering severe diabetes must inject insulin several times a day. If the insulin is not the rapid efficacy one, the patient has at risk of life. Therefore, if the storage of rapid efficacy insulin was bad and it changed to slow efficacy insulin, a serious problem will occur accompanying with risk of death. As you can see from the discussion in Fig. 2.31, rapid efficacy insulin should be stored as a fine powder. When stored as large crystals, they transform into stable crystal polymorphs. However, long ago, at a US pharmaceutical company, a large amount of rapid efficacy insulin was produced and stored in the form of large crystals for a long period, and just before shipment they grounded into fine powder and shipped it. As the results, the insulin lacked rapid efficacy, many patients died in the United States. The pharmaceutical company has paid a large amount of compensation to the bereaved families. Therefore, rapid efficacy insulin must be stored as a very fine powder from the beginning. If large crystals are stored as they are, they will rapidly change into the low efficacy stable crystal polymorphs. Pharmaceutically, fine powdering is an extremely important storage technology.

From the example of insulin above-mentioned, it can be understood that pharmaceutical companies must be well aware of not only synthesis technology of single molecule but also molecular assembly control technology.

References

[1] WC McCrone, *Polymorphism* (Chapter 8) in *Physics and Chemistry of the Organic Solid State*, Vol. II, D. Fox *et al.*, Ed., John Wiely & Sons, Int., New York (1965).
[2] J. Haleblian and W. McCrone, *J. Pharm. Sci.*, **58**, 911–929 (1969).
[3] D. Chapman, *Chem. Revs.*, **62**, 433–456 (1962).
[4] O. Mishima, *Proc. Jpn. Acad., Ser. B*, **86**, 165–175 (2010).
[5] S. Seki, H. Chihara and K. Suzuki, *Contemporary Physical Chemistry Lecture 5 Physical Chemistry of Pure Substances*, Tokyo Kagaku Dojin, Tokyo, **1967**; AS000: B000JB9754.
[6] See the DSC thermograms in Fig. 7 in Ref. [3].
[7] Example: J. Umezawa, N. Ise, R. Satoh, Y. Soeda, A. Yoshizawa, T. Kusumoto, Y. Takanishi, H. Takezoe and T. Hiyama, Anomalous Phase Transition in the Dichiral Liquid Crystalline Compound Containing Fluorines (1) Even on cooling stage, endothermic peak?, in *Proceedings of the 23rd Japanese Liquid Crystal Conference*, 3PD 06, pp. 468–469 (1997). This paper claimed a new discovery that an endothermic peak appears even on a cooling stage in the DSC measurements. However, in the discussion after this oral presentation, Professor Michio Sorai, Osaka University, stood up in the audience, and he clearly and theoretically proved by using the G–T diagram that relaxation showing an endothermic peak on cooling stage can occur in

some cases. This scene is unforgettable. The thermodynamic basis of Prof. Michio Sorai at that time is described here.

[8] (a) K. Nishikawa and I. Tanaka, *Chem. Phys. Lett.*, **244**, 149–152 (1995).

(b) K. Nishikawa, I. Tanaka and Y. Amemiya, *Chem. Phys.*, **100**, 410–421 (1996).

(c) K. Nishikawa and T. Morita, *J. Phys. Chem.*, *B*, **101**, 1413–1418 (1997).

[9] The Hayabusa 2 space probe used HMX (cyclotetramethylne tetranitramine) to excavate a hole on an asteroid, Ryugu, in 2019.

(a) T. Saiki, H. Sawada, C. Okamoto, H. Yano, Y. Takagi, Y. Akahoshi and M. Yoshikawa, *Acta Astronautica*, **84**, 227–236 (2013).

(b) https://en.wikipedia.org/wiki/Hayabusa2

[10] R. Fujishiro and A. Kuroiwa *Contemporary Physics and Chemistry Lecture 7 Nature of Solution I*, Tokyo Chemical Dojin, Tokyo (1980).

[11] M. Nonomura, K. Yamada and T. Ohta, *Research Institute of Mathematical Analysis*, **1356**, 122–127 (2004).

[12] J. W. Moore, *Moore Physical Chemistry*, 4th edition, translated by Yuichi Fujishiro, Tokyo Kagaku Doujin, Tokyo, Chapter 7(1974). Melting points of Theobroma oil: (a) https://drsuzekundu.wordpress.com/tag/chocolate/ (b)R. L. Wille and E. S. Lutton, *J. Amer. Oil Chem. Soc.*, **43**, 491–496 (1966).

[13] S. Seki and H. Suga, *Glass states of pure substances-their thermodynamics*, Chemical Review No. 5, Chem. Soc. Jpn., pp. 225 Translated by Yuichi Fujishiro,256 (1974).

Chapter 2. Exercises

1. Explain briefly what is polymorphism. Also give examples of polymorphisms.
2. How can we distinguish between polymorphs and tautomers?
3. Describe the features and differences of the first-order phase transition and the second-order phase transition.
4. Give examples of the second-order phase transition.
5. Explain monotropic and enantiotropic relationship on P–T diagram and G–T diagram.
6. Discuss relaxation.
7. Draw the HMX G–T diagram (Fig. 2.15) and explain the phase transitions of each polymorph in the heating process. Also,

explain how each polymorph forms when supercooling occurs during the cooling process.

8. (i) Explain the solubility curve and the solution phase transition.

   (ii) Why crystal polymorph having lower melting shows higher solubility?

   (iii) Explain thermodynamically why the solubility curve intersection corresponds to the solid–solid transition temperature.

9. Explain the composition phase diagrams Figs. 2.21 and 2.22.

10. The following phase transition sequence was observed in a liquid crystalline compound. Above the arrows are the phase transition temperatures and below the arrows are the phase transition enthalpy changes. K, M and L represent a crystalline phase, a liquid crystalline phase, and a liquid phase, respectively. Draw a schematic of the G–T diagram for this compound (cf. 2.2.1.2).

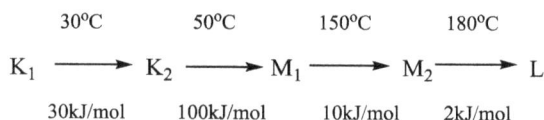

	30°C		50°C		150°C		180°C	
K_1	$\longrightarrow$	K_2	$\longrightarrow$	M_1	$\longrightarrow$	M_2	$\longrightarrow$	L
	30kJ/mol		100kJ/mol		10kJ/mol		2kJ/mol	

11. The differential scanning calorimetry (DSC) was used to determine the phase transition enthalpy change and phase transition entropy of the liquid crystalline compound PAP as follows.

(DSC measurements)

Liquid crystalline compound used for the measurement: *p*-azoxydiphenetole, abbreviation: PAP, molecular weight: 286.23

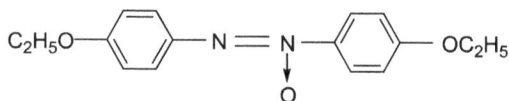

$$C_2H_5O - \!\!\!\!\bigcirc\!\!\!\! - N = N - \!\!\!\!\bigcirc\!\!\!\! - OC_2H_5$$
$$\quad\quad\quad\quad\quad\quad\;\downarrow$$
$$\quad\quad\quad\quad\quad\quad\;O$$

Reference compound: phenanthrene, molecular weight: 178.23.

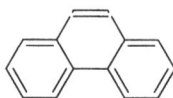

DSC measurement of 11.19 mg of the newly synthesized liquid crystalline compound PAP gave a thermogram (thermal analysis curve) as shown in B below.

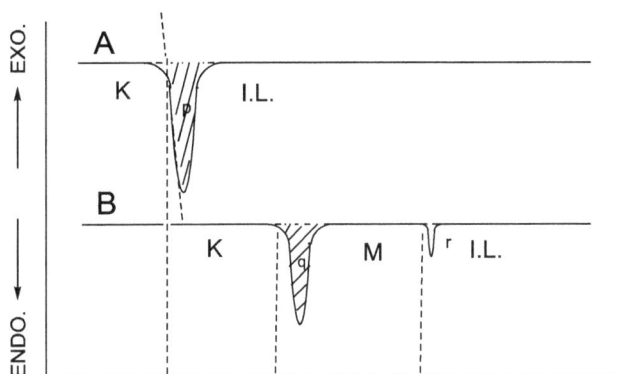

Figure 2.39. DSC thermograms.

The phase transition temperatures were determined by tangent intersection method and the following temperatures were obtained.

$$K \xrightarrow{137.8°C} M \xrightarrow{167.8°C} I.L.$$

Furthermore, since the transition enthalpy (ΔH) can also be determined from these thermogram, the following calculations were carried out.

(i) Calculation of phase transition enthalpy changes

Another DSC measurement was carried out using 18.08 mg of phenanthrene known the transition enthalpy change, under the same conditions as in the case of B. A thermogram could be obtained like A shown in the figure. The transition enthalpy change at 100°C of phenanthrene is already known to be ΔH = 4.46 kcal/mol from other methods.

In the present DSC method, the area of the hatched part p of the thermogram of A in the figure (and the weight of the paper cut off) is proportional to the enthalpy change ΔH. The weight of this paper was p = 1987 mg. The weights of

the paper cut from the thermogram of B in the figure were
q = 1125 mg and r = 86.99 mg, respectively.

By using the above data, determine the transition
enthalpy (ΔHq, ΔHr) and transition entropy (ΔSq, ΔSr)
of the liquid crystalline compound PAP.

(In the present modernized DSC measurements, the
computer program enables us automatically calculate the
phase transition enthalpies and entropies as if a black box.
However, here, in order to learn the principle, it is calculated
using the primitive method of the old days.)

(ii) Taking account of the magnitude of ΔS obtained in Q11(i),
draw a schematic of the G–T diagram for this compound
PAP (cf. Q10 and Text 2.2.1.2).

12. The liquid crystalline phase M of PAP in Question 11 is
wanted to identify by miscibility test using polarizing microscopic
observation. First, the microscopic observation revealed that this
PAP exhibited a Schlieren texture, which includes a nematic
phase and smectic C, F and I phases. Since there are two and
four brushes in the Schlieren texture of this PAP, the possibility
of the nematic phase is high. Therefore, in order to identify the
identification more surely, a miscibility test was carried out using
DSC between this unknown liquid crystalline phase-showing PAP
and the known standard nematic-phase-showing compound PAA
shown below.

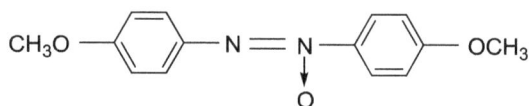

p-Azoxydianisole: abbreviation PAA; molecular
weight = 258.28

Here, if the M phase of PAP is immiscible with the standard N
phase of PAA, the binary phase diagrams may become like as
Fig. 2.40[A] shown below, and we can conclude that it is a liquid
crystal phase different from the N phase. If it mixes completely,
the binary phase diagram may become like as Fig. 2.40[B], and

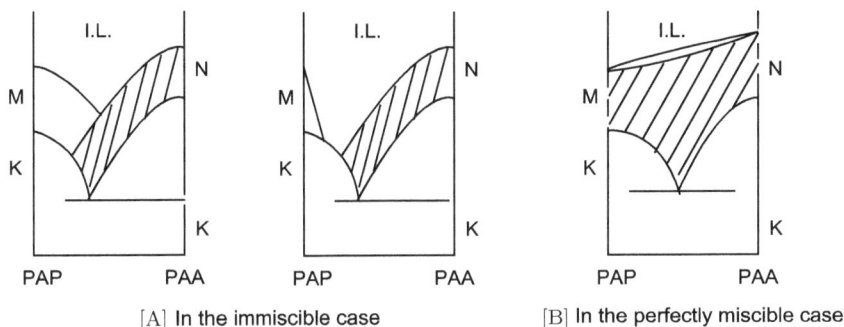

[A] In the immiscible case   [B] In the perfectly miscible case

Figure 2.40. Miscibility test.

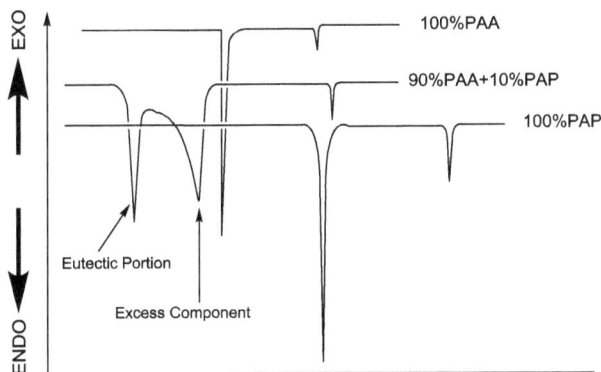

Figure 2.41. The representative DSC thermograms of the single component and the binary mixture

we can conclude that it is the same nematic phase as the N phase of PAA.

The unknown liquid crystalline phase-showing PAP was well mixed with this standard nematic-phase-showing PAA to prepare the plural DSC samples with changing the mixture ratio. The contents of PAA were listed in mol% in Table 2.2. The DSC of each sample was measured to obtain a thermogram. The representative DSC thermograms are shown in Fig. 2.41. These transition temperatures were determined by tangent intersection method and are summarized also in Table 2.2.

Using these data, draw a two-component phase diagram and show that the PAP M phase and the PAA N phase are completely

Table 2.2.   Results of miscibility test between PAA and PAP.

Mol% of PAA	Transition point (°C)	Transition point (°C)	Transition point (°C)
0	—	137.8	167.8
4.5	97.0	134.7	166.6
7.3	97.0	133.3	165.5
14.7	97.3	129.0	163.1
24.6	97.3	123.0	159.7
34.9	97.4	116.2	156.3
42.2	97.4	111.1	153.3
51.3	97.4	104.6	151.1
57.5	97.4	100.3	148.8
60.0	97.4	—	147.5
65.6	97.4	100.2	145.5
73.3	97.4	104.6	143.3
80.2	97.4	108.4	141.2
85.4	97.4	111.3	139.8
90.2	97.4	113.4	138.7
93.3	97.4	115.1	137.7
100.0	—	119.5	136.5

mixed. Also, find the experimental eutectic point from this phase diagram.

13. (i) Theoretically derive the equations of Le Chatelier–Schröder (2.9) and (2.10) in the text

   (ii) The freezing depression curves and the eutectic point can be calculated theoretically from the equations of Le Chatelier–Schröder. Draw the theoretical freezing point depression curves for PAA–PAP dealt in Questions 12 and 13, and find out the eutectic point in the theoretical drawn composition phase diagram. Verify how the theoretical ones match the experimentally observed ones.

14. Explain how to prepare metastable crystal polymorphs, taking HMX(I) to HMX(IV) as an example.

15. K (crystal) → K (crystal) phase transition is easily superheated, whereas K (crystal) → M (liquid crystal) phase transition not superheated. Why?

16. Describe the verification methods, at least two, whether polymorphs exist.

17. Explain more than three methods to prove whether two samples are crystal polymorphs of the same compound. Be sure to mention double melting behavior.
18. Discuss the application of polymorphism. Be sure to mention how to prepare suppositories and rapid efficacy insulin.
19. Read the following newspaper article and comment on this lawsuit from the viewpoint of polymorphism and drug efficacy.

December 26, 2013 Nikkei newspaper morning issue
Nissan Chemical sued for the patent infringement to her hyperlipidemia drug

On the 25th, Nissan Chemical Industries, Ltd., which manufactures the drug for hyperlipidemia treatment "Livalo," sued Tokyo District Court for the manufacturing and sales of seven generic medicines such as Towa Pharmaceutical and Daito, for infringing the Livalo patent. They are seeking a suspension of production and sales.

According to the lawsuit, these seven companies have infringed a patent on the crystalline form of the active ingredient of Livalo owned by Nissan Chemical. These seven companies began selling the corresponding generic medicines after receiving a drug price listing on March 13 because the patent for the compound was expired in August of this year. However, Nissan Chemical claims that the patent for the crystal form of this compound is still valid until 2024.

Chapter 3

Liquid Crystals: X-ray Liquid Crystal Structure Analysis

3.1 What is Mesophase?

3.1.1 *Position in three states of matter*

You may have learned in junior high school to high school that matter is composed of only three states of solid, liquid and vapor, but this idea is wrong. Recognition for new states of matter should be as shown in Fig. 3.1. In the center of this figure, a phase transition sequence from conventional solid (crystal) → liquid → vapor is shown with increasing temperature. Furthermore, in this figure, liquid crystals and plastic crystals are newly sandwiched between the solid (crystal) and the liquid. The recognition for new states of matter should be crystals, mesophases (liquid crystal and plastic crystals), liquid, vapor as shown in this figure.

Here, we should be careful about the usage of the relevant terms. The term "solid" is usually used but has some ambiguity. It is better to use "crystal" instead of "solid" Solids and crystals are not synonymous. Solid only means that it is mechanically rigid. Therefore, since ordinary inorganic glass is mechanically solid, but structurally amorphous liquid. In crystals, it is assumed that the gravity center positions and orientations of molecules (atoms) are regularly arranged in three dimensions infinitely. Since inorganic glass, which is commonly used for windows and others, has no

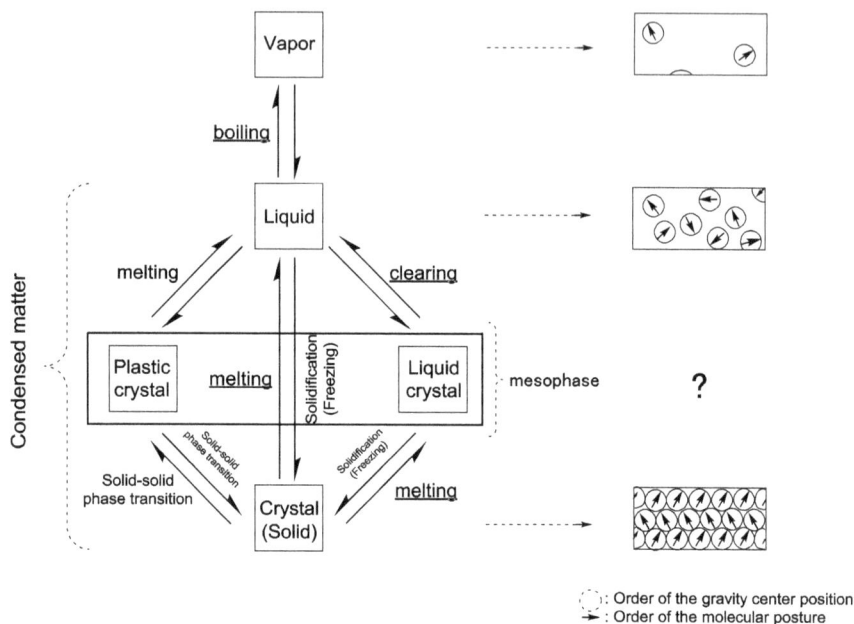

Figure 3.1.   Novel recognition of states of matter.

positional order of atoms, it is solid but not crystallographically crystal. So, academically, it would be better that the term solid in three states of matter is replaced by the term crystal.

We should also be very careful about the definition of glass. It is generally recognized that glass used for windows and others is an inorganic compound and glass can be obtained by quenching molten alkali silicate or the like. However, the academic definition of glass is a state showing glass transition point. Glass is a state where some disorderliness is frozen as it is, below the glass transition point. Since such a glass transition point can be seen for not only inorganic compounds but also organic compounds and polymers, most of the materials become glass. When even water and ethyl alcohol are rapidly cooled, they become glass. Since disorderliness of the state before quenching is frozen, we can obtain glassy liquid, glassy liquid crystal, glassy crystal, etc., depending on the state before quenching.

The detailed excellent basic research and definition of glass were first done by Seki and his coworkers in Osaka University. Therefore, for details of glass, please refer to their review [1].

As you can see around the centerline of Fig. 3.1, with heating crystals melt into liquid, and then liquid boils into a vapor, according to the traditional concept of three states of matter. In this figure, liquid crystal and plastic crystal are additionally sandwiched between the crystal and the liquid. Conventional states of matter does not include these states. Liquid crystals and plastic crystals between crystals and liquid are collectively referred to as mesophase. This is the new recognition of the states of matter: crystal, mesophases (liquid crystal and plastic crystals), liquid and vapor, as illustrated in this figure.

Therefore, we have to rename the terms of phase transitions. In the conventional definition, melting means the phase transition from crystals to liquid. However, when the mesophase intervenes between crystals and liquid, it should be defined as follows: melting means the phase transition from crystal to liquid crystal, so that the phase transition from liquid crystal to liquid is newly defined as clearing. On the other hand, the phase transition from crystal to plastic crystal is regarded as a phase transition from a crystal to another crystal transition, so that the phase transition from plastic crystal to liquid is defined as melting.

On the right side of this figure, the gravity center position of the molecules is indicated by a circle, the molecular posture (orientation) is indicated by an arrow, and the phase structures of the molecular assembly states are schematically illustrated. As can be seen from these schematic diagrams, in the **crystal**, the gravity center position and orientation of the molecules are regularly arranged three-dimensionally. However, when the crystal is heated up to liquid, in the **liquid**, the gravity center position and orientation of the molecules are randomized. When the liquid is further heated up to vapor, in the **vapor**, they are much more randomized. So, what kinds of phase structures have the mesophases (liquid crystal and plastic crystal)? Next we will consider the characteristics of the molecules exhibiting these mesophases and their structures.

3.1.2 *Plastic crystal*

The concept of plastic crystals was submitted by Timmermans in 1938.

3.1.2.1 *Features*

The plastic crystal phase is found in the crystal phase of carbon tetrachloride, cyclohexanol and the like, just below the melting point. Since the plastic crystalline phase gives X-ray Laue spots and exhibits a clear melting point, it satisfies the definition as a crystal. However, from a mechanical viewpoint, it is far from the hardness possessed by ordinary crystals and has remarkable plasticity. In the outstanding cases, plastic flow occurs by its own weight.

3.1.2.2 *Judgment conditions*

(1) The molecular structure is nearly spherical.
(2) The melting entropy ΔS should be small.

As the upper limit, about $21 \text{J/K} \cdot \text{mol}$ = Empirical rule of Timmermans.
The fact that the melting entropy ΔS of is small means that the order of the plastic crystal does not differ much from the order of the liquid. That is, a plastic crystal is a crystal close to a liquid. Although close to liquid, this X-ray diffraction pattern is classified as a crystalline phase because it indicates the presence of a long-distance three-dimensional space lattice such as a crystal.

3.1.2.3 *Examples of plastic crystals (See (1) and (2) in the judgment conditions.)*

Figure 3.2 illustrates examples of the compounds that show a plastic crystal phase. As can be seen from this figure, each of the compounds has a nearly spherical molecular structure:

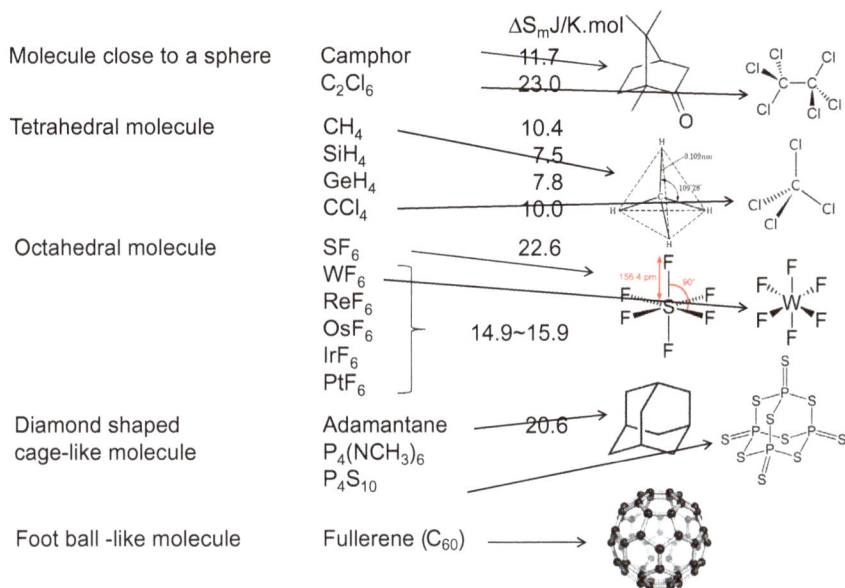

Figure 3.2. Examples of plastic crystal.

tetrahedral molecules such as methane and carbon tetrachloride, **octahedral molecules** such as tungsten hexafluoride, **diamond-cage-shaped** molecules such as adamantane and phosphorus pentasulfide, and **soccerball-shaped molecules** such as fullerenes. Each of the molecules exhibiting a plastic crystal phases is not perfect sphere but close to sphere. Furthermore, as can be seen from this figure, most of the melting entropies ΔSm of the phase transition from crystals to plastic crystals are smaller than 21 J/K·mol, which is in agreement with Empirical rule of Timmermans.

3.1.3 *Liquid crystal*

In 1888, F. Reinitzer first recognized it as a matter-specific phenomenon, and in 1889 O. Lehmann named it liquid crystal.

3.1.3.1 *Features*

When a crystal is heated, it melt first at T_1 (K) to form a turbid viscous liquid, and then it turns to a clear liquid at $T_2 (> T_1)$.

The liquid formed between T_1 and T_2 exhibits birefringence (optical anisotropy) under a polarization microscope. Birefringence is characteristic to crystal and fluidity is a characteristic to liquid. Therefore, this state was named liquid crystal from the meaning of crystal showing fluidity.

In liquid crystals, there is no long distance three-dimensional space lattice like crystals. Reduction of the long distance order of the molecular gravity center causes to show plasticity or fluidity. However, since partial molecular orientation remains, it exhibits optical anisotropy. The simultaneous appearance of both fluidity (or plasticity) and optical anisotropy is a major feature of liquid crystals.

(Caution 1) Some liquid crystal phases do not spontaneously flow. These are soft and flexible when pressed. Therefore, not all the liquid crystal phases have fluidity.

(Caution 2) In the "hard" liquid crystal phase (from S_G to S_L), a very short distance three-dimensional space lattice exists. These liquid crystal phases have long been an endless debated as to whether they are "hard" liquid crystal phases or "soft" crystal phases. They are the phases in the gray zone between the crystal and the liquid crystal.

3.1.3.2 *Monotropic liquid crystal and enantiotropic liquid crystal*

Figure 3.3 shows the phase transition sequences and the G–T diagram for [A] enantiotropic liquid crystal phase and [B] monotropic liquid crystal phase. They correspond to the [A] enantiotropic relationship and the [B] monotropic relationship of crystal polymorphs I and II in Fig. 2.10 shown in Chapter 2, respectively. Here, the relationship between the crystal phase K and the liquid crystal phase L.C. is [A] enantiotropic and [B] monotropic.

3.1.3.3 *Appearance of the states of matter*

Figure 3.4 schematically illustrates the appearance of the states of matter: [A] crystal having an own shape like needle, plate and

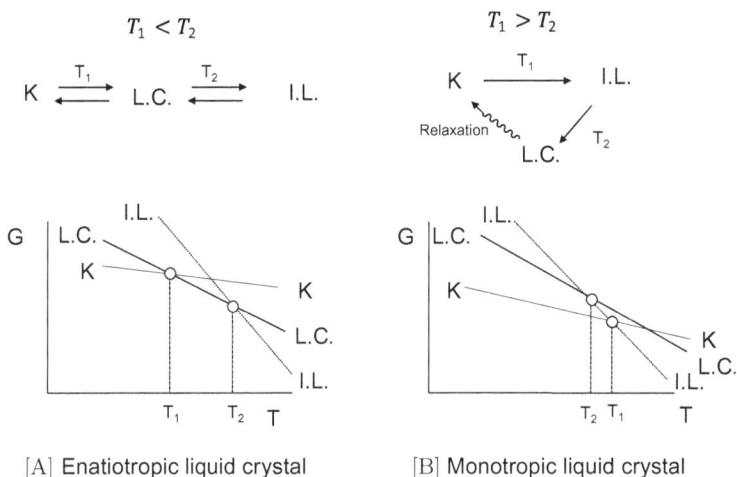

[A] Enatiotropic liquid crystal   [B] Monotropic liquid crystal

Figure 3.3. Phase transition sequences and G-T diagrams for [A] enantiotropic liquid crystalline phase and [B] monotropic liquid crystalline phase.

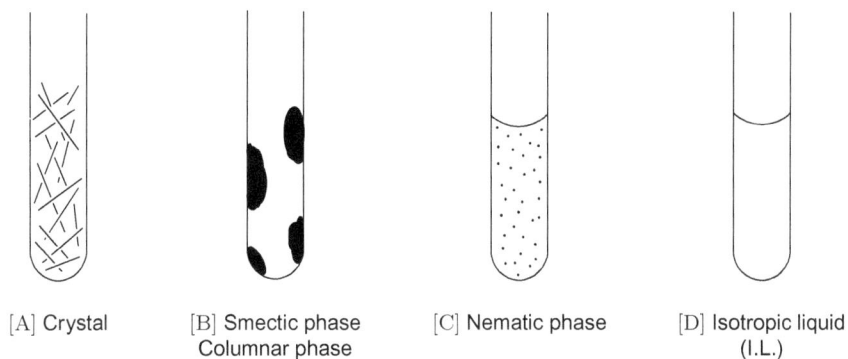

[A] Crystal   [B] Smectic phase   [C] Nematic phase   [D] Isotropic liquid
                 Columnar phase   (I.L.)

Figure 3.4. Appearance of the states of matter: [A] crystal having an own shape like needle, plate and the others; [B] smectic phases and the columnar phase like a paste; [C] nematic phase like a cloudy liquid. It shows spontaneous fluidity and meniscus; [D] clear isotropic liquid showing fluidity and meniscus.

the others; [B] smectic phases and the columnar phase like a paste; [C] nematic phase like a cloudy liquid. It shows spontaneous fluidity and meniscus; [D] clear isotropic liquid showing fluidity and meniscus.

We should pay attention that there are also many liquid crystal phases having appearance [B]. It is because that many textbooks of liquid crystals have written that liquid crystal phases have (spontaneous) fluidity and optical anisotropy. Accordingly, if you expect only the appearance [C], you cannot find out the liquid crystal phases having the appearance [B] When you encounter a liquid crystal phase having the appearance [B], it is better to press the cover glass from above to check whether it is soft or not.

As can be seen from these appearances, crystals have their own specific shape, but liquid crystals and liquids do not have their own specific shape. Therefore, when performing the X-ray diffraction experiments, a sample of crystal can be used by bonding the single crystal to the tip of a capillary, but a sample of liquid crystal and liquid needs a container to fill in. Therefore, you need a special device for setting the sample of liquid crystal and liquid when performing the X-ray diffraction experiments (cf. Section 3.5.7).

3.1.3.4 *Judgment conditions*

(1) Liquid crystals have a rod-like or disc-like molecular structure and long alkyl chains in the periphery (cf. Section 3.1.3.5)
(2) It shows both birefringence and fluidity (or plasticity).
(3) The X-ray diffraction pattern does not indicate the presence of a long distance three-dimensional space lattice.

(Note 1) Generally, when the liquid crystal is heated, the long-chain alkyl groups in the periphery melt first to form a soft liquid part, and the central core does not melt, forming a rigid crystal part, so it is thought that a liquid crystal phase appears as a whole. Therefore, it has been believed for over 100 years that a molecule should have long- alkyl chains indispensible for the appearance of the liquid crystal phase. However, recently, it has been revealed that bulky substituents can also induce liquid crystal phases instead of long-chain alkyl groups (flying-seed-like liquid crystals) [2].

(Caution 2) When the liquid crystal phase is observed in a micro scale, a soft liquid portion is formed in the periphery and a rigid crystal portion is formed in the central core, so it is considered that the liquid crystal phase appears as a whole. However, the liquid crystal phase is by no means a dispersion system: you should not regard that fine crystals are mixed and dispersed in liquid and in a micro scale (i.e., two phase separation system). It should be noted that the liquid crystal phase is one thermodynamically independent phase. Hence, the term of microphase separation should never be used for liquid crystal phases (cf. Section 2.3.4.3 in Chapter 2).

3.1.3.5 *Examples of liquid crystals*

Figure 3.5 shows examples of liquid crystalline compounds. Entries [A] and [B] are **calamitic liquid crystals** having a rod-like molecular structure, and entries [C] to [E] are **discotic liquid crystals** having a disc-like molecular structure. In addition, entries [A] to [D] are liquid crystals having long alkyl chain groups at the periphery, and entry [E] is **flying-seed-like liquid crystals** having bulky groups instead of the long alkyl chain groups.

Since the first liquid crystal compound was discovered by Reinitzer in 1888, over 100,000 liquid crystal compounds have been synthesized to date [3].

3.1.4 *Molecular arrangements in mesomorphic phases (plastic crystal and liquid crystal)*

Figure 3.6 schematically shows how the molecules of plastic crystals and liquid crystals are arranged.

Figure 3.6[A] represents what arrangements of the nearly spherical molecules change from crystal phase to plastic crystal phase. In this figure, molecules close to a spherical shape are represented by elliptical balls such as American football (= rugbyball). In this crystal state [A](1), the gravity center of the ball is in a body-centered cubic lattice, and the posture of the ball indicated by the arrows is also regularly arranged. Therefore, in this crystalline state, there are both the three-dimensional order of the gravity center position and the order of the posture. The molecule is so close to a sphere that the rotational barrier is small. Accordingly,

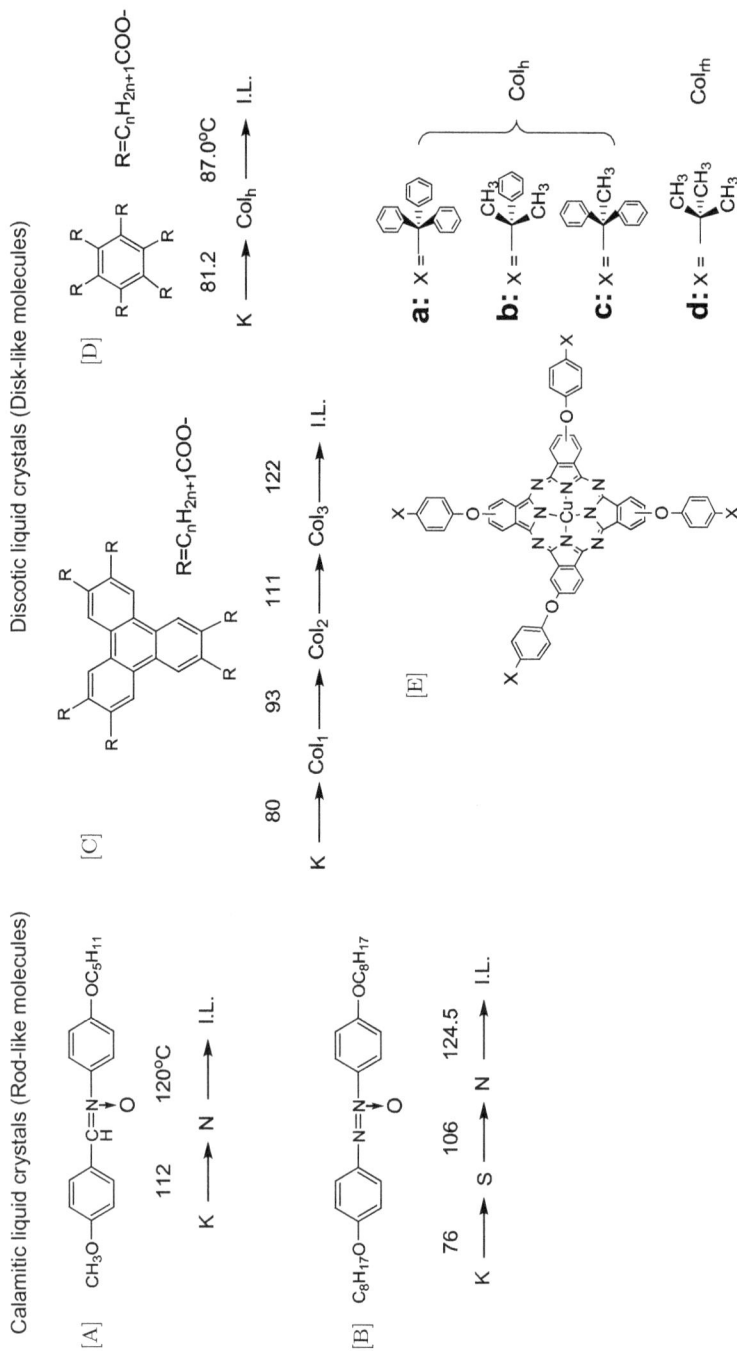

Figure 3.5.　Examples of liquid crystals.

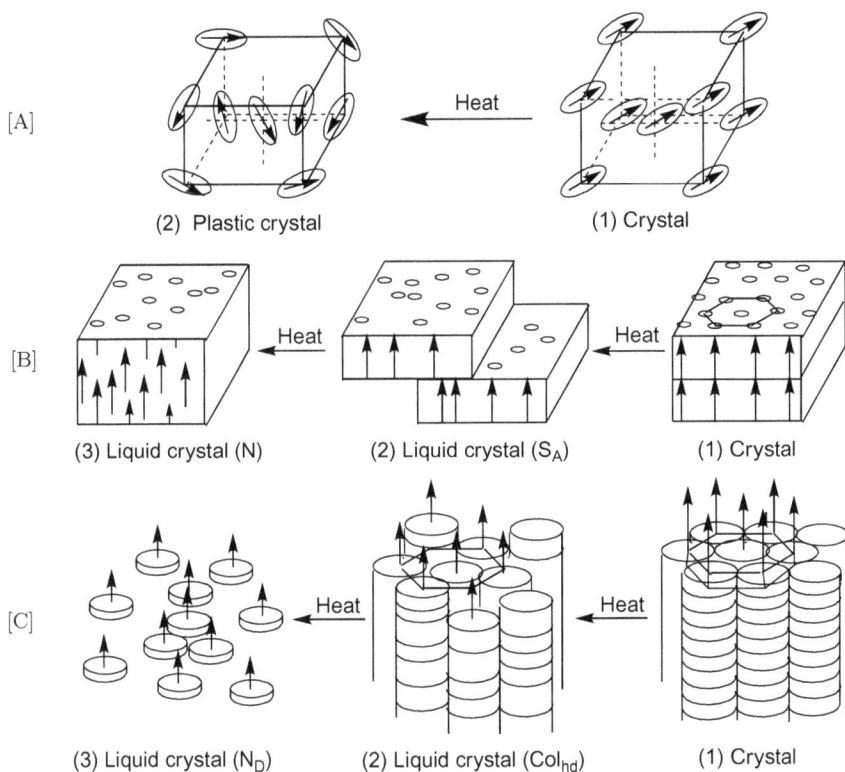

[A]

(2) Plastic crystal (1) Crystal

[B]

(3) Liquid crystal (N) (2) Liquid crystal (S_A) (1) Crystal

[C]

(3) Liquid crystal (N_D) (2) Liquid crystal (Col_{hd}) (1) Crystal

Figure 3.6. Molecular arrangements for plastic crystal and liquid crystal.

when this crystal is heated, the molecule freely rotates to break the order of the molecular posture at first. However, the gravity center position of the molecules remains in a body-centered cubic lattice, as it is. Such a state is a plastic crystal [A](2).

Figure 3.6[B] represents what arrangements of rod-like molecules change from crystal [B](1) to smectic A (S_A) phase [B](2) and further transition to nematic (N) phase [B](3). In this figure, rod-like molecules are simply represented by arrows. In this crystal state [B](1), the arrows form a three-dimensional hexagonal lattice, and the orientations of the arrows are regularly arranged to form a layer structure. Therefore, in this crystalline state, there are both the three-dimensional order of the gravity center position and the order of the posture. The intermolecular force in the rod-like molecule is large side by side because the contact area is large in the lateral direction On the other hand, the intermolecular force is small between the head and tail of the molecules, because the contact area of the molecule is small in the longitudinal direction, Therefore, when this crystal is heated, it starts to slip between the weak intermolecular forces. In this figure [B](2), the sliding of the layers is illustrated by the displacement of the layers. Also, in the smectic A (S_A) phase, the order of the two-dimensional hexagonal lattice in the layer is broken. Therefore, it is remained in this S_A phase that the order of the one-dimensional gravity center position in the layer direction and the order of the molecular posture. When this S_A phase is further heated, the strong intermolecular force of the rod-like molecules in the lateral direction is finally overcome. Accordingly, as shown in figure [B](3), the molecules slide up and down and the layer structure is broken. Nevertheless, the molecular posture still remains in a certain direction. The phase structure is a nematic liquid crystal (N) phase.

Figure 3.6[C] represents what arrangements of disk-like molecules change from crystal [C](1) to hexagonal disordered columnar (Col_{hd}) phase [C](2) and further to discotic nematic (N_D) phase [C](3). In this figure, the molecular posture of disk-like molecules is represented by arrows. In this crystal state [C](1), the arrows form a three-dimensional hexagonal lattice, and the postures of the arrows are regularly arranged in a direction. Therefore, in this crystalline state, there are both the three-dimensional order of the gravity center position and the order of the posture. Since disk-like molecules have small contact area in the lateral direction they have small intermolecular force in this direction. On the other hand, the face-to-face contact area of molecules is so large that intermolecular force is very big in the longitudinal direction of the columnar structure. Therefore, when this crystal is heated, at first the columns begin to slip, as shown in figure [C](2). This figure illustrates the sliding columns by the height difference of these columns. There are still two-dimensional hexagonal lattice among the columns. Also, the stacking order of disks in the column is broken because the distance between the face-to-face molecules is thermally fluctuated. Therefore, in this disordered columnar hexagonal (Col$_{hd}$) phase a two-dimensional hexagonal lattice is formed among the surrounding columns and it is remained the order of the two-dimensional gravity center and the order of the molecular posture. When this Col$_{hd}$ phase is further heated, the strong intermolecular force in the longitudinal direction of the disk-like molecules is overcome to slip the disks left and right in between; the column structure is finally broken with keeping the molecular posture in one direction, as shown in figure [C](3) This phase structure is a discotic nematic (N$_D$) liquid crystal phase. When you look only at these arrows, you can easily recognize that the N$_D$ phase of disk-like molecules in [C](3) has the same phase structure of the N phase of rod-like molecules in [B](3).

3.1.5 *Summary of features of mesophases*

Melting from (ordered) crystals to (isotropic) liquids is thought to have two aspects: gradual disintegration of the three-dimensional crystal lattice and disintegration of the molecular orientation (posture). If the molecules are close to spherical and the barrier to intramolecular crystal rotation is small, a disintegration of molecular orientation is realized at first with maintaining the three-dimensional crystal lattice, to give a plastic crystal. On the contrary, in rod-like molecules and discotic molecules, if the intermolecular forces to align parallel or perpendicular to each other are strong, stepwise disintegration of the three-dimensional crystal lattice occurs first to give liquid crystals with maintaining the molecular orientation. Thus, the two intermediate states are resulted by the separate disintegration of the two orders and are complementary to each other. If this is put together in a table, it will become like Table 3.1.

In this table, the "long-distance 3D order of gravity center position" of liquid crystal occurs stepwise disintegration, which is denoted as $\Delta \rightarrow \times$. Most of the conventional textbooks for liquid

Table 3.1. Features of the mesomorphic states: the difference between plastic crystal and liquid crystal.

Order / State	Long-distance 3D order of gravity center position	Long-distance 3D order of molecular posture
Liquid	✗	✗
Plastic crystal	◯	✗
Liquid crystal	✗ ←Stepwise disintegration— △	◯
Crystal	◯	◯

Intermediate states (mesophases): Plastic crystal, Liquid crystal

X: this mark means that the order does NOT exist.   O: this mark means that the order exists.

crystal have long written that there is no "order of long-distance gravity center position" in liquid crystals. If so, we should indicate liquid crystalline state as zero dimension represented by x in this table. However, it corresponds only to a nematic phase. As seen above, the liquid crystalline state includes many smectic phases and columnar phases in which dimensionality still remains such as 1D and 2D. Hence, the conventional concept is wrong. For details of the dimensionality of various liquid crystal phases, see Section 3.3 and Ref [4].

3.2   Types of Liquid Crystal

There are several types of liquid crystal classification. Two representative ones are listed below. These classifications are summarized in Table 3.2.

3.2.1   *Thermotropic liquid crystal and lyotropic liquid crystal*

The thermotropic liquid crystal means the compound that shows crystalline states at lower temperature but liquid crystalline states at higher temperatures.

Thermotropic liquid crystals are further divided into **calamitic liquid crystals** represented by rod-like molecules and **discotic**

Table 3.2. Types of liquid crystals.

Thermotropic Liquid Crystals: The material that is crystalline states at lower temperatures but liquid crystalline states at higher temperatures.

Lyotropic Liquid Crystals: The material that is crystalline states as it is, but liquid crystalline states for the solution in some concentration.

Thermotropic Liquid Crystals

Calamitic Liquid Crystals (Rod-like molecules)		Smectic phases (S_{A-C}, S_{E-L}, S_T: 12 kinds) Nematic phases Cholesteric phases Cubic phases
Discotic Liquid Crystals (Disk-like molecules)		Columnar phases (Col_h, Col, Col_{tet}, Col_{ob}) Discotic nematic phase Discotic cholesteric phase Discotic lamellar phases Cubic phases
Flying-seed-like Liquid Crystals		
Lyotropic Liquid Crystals		Middle phases Neat phase Cubic phases

Low molecular liquid crystals
Polymer liquid crystals

liquid crystals represented by disk-like molecules. Furthermore, recently, there is a new type of **flying-seed-like liquid crystals** which has a bulky substituent instead of a long chain alkyl group (cf. Fig. 3.5[E]). The calamitic liquid crystals are further subdivided into smectic phases, nematic phases, cholesteric phases and cubic phases depending on the difference in the phase structure. In addition, the discotic liquid crystals are further subdivided into columnar liquid crystal phases, discotic nematic phases, discotic cholesteric phases, discotic lamellar phases and cubic phases depending on the difference in the phase structure. Thus, thermotropic liquid crystals can be subdivided into many types according to the difference in their molecular structure and phase structure.

On the other hand, lyotropic liquid crystal means that the liquid crystal states appear depending on the concentration conditions when dissolved in a suitable solvent. A lot of people may have experienced the case where the soap is soaked with water to swollen and soften in the bathroom. Such a state is lyotropic liquid crystal. In addition, the cytoplasmic flow of cells is also in the lyotropic liquid crystalline state, and a 60% concentrated aqueous solution of DNA is also in the liquid crystalline state. Thus, the organism is closely related to the lyotropic liquid crystalline state. Silkworm and spider discharges the solution in the lyotropic liquid crystal state from the mouth to produce silk and cocoon threads. This is a spinning mechanism called liquid crystal spinning. Kevlar fibers used in bulletproof vests are fabricated by this liquid crystal spinning.

As can be seen from Table 3.2, the lyotropic liquid crystals are also subdivided into middle phases (= hexagonal columnar phases), neat phases (= lamellar (smectic) phases), and cubic phases, depending on the difference in their phase structures.

Although the liquid crystal phases of the thermotropic liquid crystal and the lyotropic liquid crystal have similar phase structures, different names are used because the researchers were originally different from each other and there has had no good exchange of knowledge between them. It is hoped that in the future, the same name will be used for the same phase structure.

3.2.2   *Low molecular weight liquid crystal and polymer liquid crystal*

This classification is mainly made from differences in industrial applications. **Low molecular weight liquid crystals** are currently widely used as display materials represented by liquid crystal televisions. Liquid crystal materials used in liquid crystal televisions are low molecular dielectric materials exhibiting a nematic liquid crystal phase. On the other hand, **polymer liquid crystals** are widely used as materials for engineering plastics with high dimensional accuracy and high strength fibers.

3.3   Dimensionality and Hierarchy of Mesophases

As described above, there are various liquid crystal phases, so that their molecular assembly structures are also various. Therefore, from now on, we will discuss about these various liquid crystal phase structures and the X-ray structure analysis methods of those liquid crystal phases.

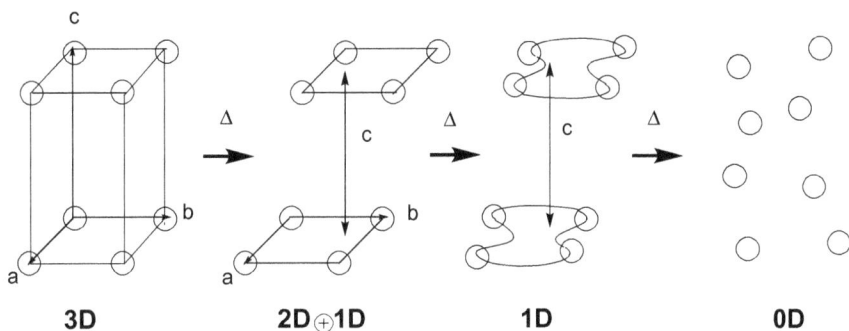

Figure 3.7.   Schematic illustration of stepwise lattice disintegration of a crystal by heating or adding solvent.

3.3.1   *Step-wise disintegration of crystal lattice*

First, we mathematically consider how liquid crystals form between crystals and liquid in the process of disintegration of the three-dimensional crystal lattices. Figure 3.7 schematically depicts how a three-dimensional (3D) crystal lattice disintegrates stepwise into a zero-dimensional (0D) isotropic liquid by heating or adding solvent. In this figure, an orthorhombic crystal lattice is illustrated as an example. When this 3D orthorhombic crystal is heated, the upper and lower bases slide out. It is because that the distance in *c*-axis direction is the longest and so the intermolecular force in this direction is the weakest. Therefore, it is divided into a two-dimensional (2D) *ab* lattice and a one-dimensional (1D) *c* lattice. This state is mathematically expressed as 2D⊕1D using the symbol ⊕ of the direct sum. 2D⊕1D is never a 3D space. This is a direct sum of a two-dimensional vector subspace and a one-dimensional vector subspace, which means that the subspaces are perpendicular to each other. When this [2D⊕1D] liquid crystal phase is further heated, the two-dimensional *ab* lattice disintegrates, but the layer structure is still maintained in the *c*-axis direction. Therefore, this liquid crystal phase is 1D. When heated furthermore, the molecules can move up and down and the layer structure finally disintegrates. At this time, there is no longer any dimensionality, and this state is zero-dimensionality (0D) In the case where the order of the molecular posture is also broken in this zero-dimensional molecular assembly, it corresponds to an isotropic liquid (I.L.). Isotropic liquids (IL) are generally referred to as liquids.

Next, from mathematical expressions we will further consider these stepwise disintegration of crystal lattice that a 3D crystal gradually breaks down into a [2D⊕1D] and a 1D liquid crystal phases.

3.3.1.1 *Lattice disintegration of orthorhombic crystal system*

First, an orthorhombic crystal system will be described as an example. The relationship among the spacing d, the Miller index and the lattice constants in the orthorhombic 3D crystal lattice is shown in Table 3.3 From this table,

$$(3D): \frac{1}{d_{hkl}^2} = \frac{h^2}{a^2} + \frac{k^2}{b^2} + \frac{l^2}{c^2} \tag{3.1}$$

From this 3D crystal lattice, we can observe all the X-ray reflections of three-dimensional $(hk\ell)$, two-dimensional $(hk0)$, and one-dimensional (00ℓ) reflection lines, as shown in Table 3.4. However, when heated this 3D crystal lattice, the ab planes between the upper and lower planes begin to slip, and the correlation between these ab planes loses; the long-distance 3D order of gravity center order disappears. At this time, the original 3D lattice disintegrations into a 2D lattice and a 1D lattice as shown in Fig. 3.7. In that case, the relational expression of (3.1) is divided into following two expressions of Equation (3.2.1) and Equation (3.2.2),

$$(2D): \frac{1}{d_{hk0}^2} = \frac{h^2}{a^2} + \frac{k^2}{b^2} \tag{3.2.1}$$

$$(1D): \frac{1}{d_{00l}^2} = \frac{l^2}{c^2} \tag{3.2.2}$$

At this time, 2D $(hk0)$ reflection and 1D (00ℓ) reflection are observed as the X-ray reflection, and 3D $(hk\ell)$ reflections are no longer observed. This state corresponds to the smectic E (S_E)

Table 3.3. Relationship among spacing (d), Miller indices (hkl) and the lattice constants $(a, b, c, \alpha, \beta$ and $\gamma)$ for [2D $\oplus$ 1D] mesophases known to date.

3D	2D $\oplus$ 1D	Lattices after degradation	Mesophase
Orthorhombic $$\frac{1}{d_{hkl}^2} = \frac{h^2}{a^2} + \frac{k^2}{b^2} + \frac{l^2}{c^2}$$	$\Rightarrow$ $$\frac{1}{d_{hk0}^2} = \frac{h^2}{a^2} + \frac{k^2}{b^2}$$ $$\frac{1}{d_{00l}^2} = \frac{l^2}{c^2}$$	Rectangular Stacking distance(c = h), Layer thickness (c)	Col$_{ro}$ (a>b>>c), S$_E$ (c>>a>b)
3D-hexagonal $$\frac{1}{d_{hkl}^2} = \frac{4}{3}\left(\frac{h^2+hk+k^2}{a^2}\right) + \frac{l^2}{c^2}$$	$\Rightarrow$ $$\frac{1}{d_{hk0}^2} = \frac{4}{3}\left(\frac{h^2+hk+k^2}{a^2}\right)$$ $$\frac{1}{d_{00l}^2} = \frac{l^2}{c^2}$$	2D-hexagonal Stacking distance (c = h), Layer thickness (c)	Col$_{ho}$ (a>>c), S$_B$ (c>>a) (S$_L$)
3D-tetragonal $$\frac{1}{d_{hkl}^2} = \frac{h^2 + k^2}{a^2} + \frac{l^2}{c^2}$$	$\Rightarrow$ $$\frac{1}{d_{hk0}^2} = \frac{h^2 + k^2}{a^2}$$ $$\frac{1}{d_{00l}^2} = \frac{l^2}{c^2}$$	2D-tetragonal Stacking distance (c = h), Layer thickness (c)	Col$_{tet.o}$ (a>>c), S$_T$ (c>>a)
Monoclinic $\alpha = 90°, \beta = 90°, \gamma \neq 90°$ a>b>>c $$\frac{1}{d_{hkl}^2} = \frac{1}{\sin^2\gamma}\left(\frac{h^2}{a^2} + \frac{k^2}{b^2} - \frac{2hk\cos\gamma}{ab}\right) + \frac{l^2}{c^2}$$	The a axis tilts to the b axis direction. $\Rightarrow$ $$\frac{1}{d_{hk0}^2} = \frac{1}{\sin^2\gamma}\left(\frac{h^2}{a^2} + \frac{k^2}{b^2} - \frac{2hk\cos\gamma}{ab}\right)$$ $$\frac{1}{d_{00l}^2} = \frac{l^2}{c^2}$$	Oblique (a>b>>c) $\gamma \neq 90°$ Stacking distance (c = h)	Col$_{ob.o}$
Monoclinic $\alpha = 90°, \beta \neq 90°, \gamma = 90°$ c>>a>b $$\frac{1}{d_{hkl}^2} = \frac{1}{\sin^2\beta}\left(\frac{h^2}{a^2} + \frac{l^2}{c^2} - \frac{2hl\cos\beta}{ac}\right) + \frac{k^2}{b^2}$$	The c axis tilts to the a axis direction. $\Rightarrow$ $$\frac{1}{d_{hk0}^2} = \frac{h^2}{a^2} + \frac{k^2}{b^2}$$ $$\frac{1}{d_{00l}^2} = \frac{l^2}{c^2}$$	Rectangular (c>>a>b) $\beta \neq 90°$ Layer thickness (c)	S$_F$ (S$_G$) (S$_H$)
Monoclinic $\alpha \neq 90°, \beta = 90°, \gamma = 90°$ c>>a>b $$\frac{1}{d_{hkl}^2} = \frac{1}{\sin^2\alpha}\left(\frac{k^2}{b^2} + \frac{l^2}{c^2} - \frac{2kl\cos\alpha}{bc}\right) + \frac{h^2}{a^2}$$	The c axis tilts to the b axis direction. $\Rightarrow$ $$\frac{1}{d_{hk0}^2} = \frac{h^2}{a^2} + \frac{k^2}{b^2}$$ $$\frac{1}{d_{00l}^2} = \frac{l^2}{c^2}$$	Rectangular (c>>a>b) $\alpha \neq 90°$ Layer thickness (c)	S$_I$ (S$_J$) (S$_K$)

d: spacing; h, k, l: Miller index; a, b, c, α, β, γ: lattice constants.

Definition: α is the angle between b and c axes; β is the angle between a and c axes; γ is the angle between a and b axes.

phase, the discotic ordered rectangular columnar (Col$_{ro}$) phase, and the discotic lamellar rectangular (D$_{L.rec}$) phase (see Table 3.6 in Section 3.3.3). On further heating, the 2D lattice distorts to lose the long-distance 2D order of gravity center order. At this time, the original 2D lattice disintegrates, leaving only the 1D lattice. In that case, only Equation 2.2 survives, and only 1D (00ℓ) reflections are observed for the X-ray reflections This state corresponds to the cases of the S$_A$, S$_C$ and D$_{L1}$ phase. At this time, we will consider Equation (3.2.2). When extracted the square root of both sides of

Table 3.4.   Relationship between phase dimensionality and X-ray reflection line dimensionality.

Reflection line dimensionality	Phase dimensionality						
	3D	[2D⊕1D]	2D	1D	0D	[1D⊕1D]	[1D⊕1D ⊕1D]
1-dimensional	(00l)	(00l)	—	(00l)	—	(h00)+(00l)	(h00)+(0k0)+(00l)
2-dimensional	(hk0)	(hk0)	(hk0)	—	—	—	—
3-dimensional	(hkl)	—	—	—	—	—	—
Single(S) or composite (C) lattice-based phase	S	C	S	S	S	C	C

this equation, since $(\ell, c \geq 0)$, so

$$\frac{1}{d_{00l}} = \frac{l}{c} \qquad (3.2.2.1)$$

Since the lattice constant c in this equation is constant,

$$d_{00l} \propto \frac{1}{l} \qquad (3.2.2.2)$$

Therefore, the spacing $d_{00\ell}$ and the Miller index ℓ are inversely proportional. Since $\ell = 1, 2, 3, 4, 5, \dots$ the following spacing ratios can be obtained:

$$d_{001} : d_{002} : d_{003} : d_{004} : d_{005} : \cdots = 1 : \frac{1}{2} : \frac{1}{3} : \frac{1}{4} : \frac{1}{5} : \cdots \qquad (3.3)$$

Such ratios are observed in a lamellar liquid crystal phase like as smectic phases or discotic lamellar phases.

When the 1D lattice of Equation (3.2) is further heated, this 1D lattice also disintegrates. At this time, the ratios in Equation (3.3) for the 1D (00ℓ) reflections are no longer observed, and only a broad reflection due to the 0D nature is given. These states correspond to the case of nematic phase or isotropic liquid.

3.3.1.2 *Lattice disintegration of hexagonal crystal system*

Next, the hexagonal crystal system is considered. As can be seen from Table 3.3, the relationship among the spacing d, the Miller index and lattice constant in the hexagonal 3D crystal lattice is

$$(3D) : \frac{1}{d_{hkl}^2} = \frac{4}{3} \left(\frac{h^2 + hk + k^2}{a^2} \right) + \frac{l^2}{c^2} \qquad (3.4)$$

Similarly to the previous example, when heated, the original 3D lattice disintegrates into a 2D lattice and a 1D lattice. So, expression of (3.4) is divided into the following two expressions:

$$(2D) : \frac{1}{d_{hk0}^2} = \frac{4}{3}\left(\frac{h^2 + hk + k^2}{a^2}\right) \qquad (3.4.1)$$

$$(1D) : \frac{1}{d_{00l}^2} = \frac{l^2}{c^2} \qquad (3.4.2)$$

From Equation (3.4.2), the same relationship as Equation (3.3) is obtained.

The (00ℓ) reflection obtained from Equation (3.4.2) corresponds to the lamellar structure in the S_B phase and the stacking distance in the column in the Col_{ho} phase.

When extracted the square root of Equation (3.4.1) and organized.

$$d_{hk0} = \frac{\sqrt{3}a}{2\sqrt{h^2 + hk + k^2}} \qquad (3.4.1.1)$$

Since the lattice constant a in this equation is constant,

$$d_{hk0} \propto \frac{1}{\sqrt{h^2 + hk + k^2}} \qquad (3.4.1.2)$$

When $(hk) = (10), (11), (20), (21), (30), (22), \ldots$, following ratios are obtained:

$$d_{100} : d_{110} : d_{200} : d_{210} : d_{300} : d_{220} : \cdots$$
$$= 1 : \frac{1}{\sqrt{3}} : \frac{1}{2} : \frac{1}{\sqrt{7}} : \frac{1}{3} : \frac{1}{\sqrt{12}} : \cdots \qquad (3.5)$$

Comparing these present ratios in Equation (3.5) with the previous ratios in Equation (3.3), the characteristic reflections having the ratios such as $\frac{1}{\sqrt{3}}$, $\frac{1}{\sqrt{7}}$ and $\frac{1}{\sqrt{12}}$ are observed. This state corresponds to the cases of the S_B and Col_h phases. The characteristic reflections serve as a good guideline for the identification as the S_B and Col_h phases.

3.3.1.3 *Lattice disintegration of tetragonal crystal system*

Furthermore, the tetragonal crystal system is considered in the same manner as the above examples. As can be seen from Table 3.3, the following 2D-tetragonal lattice can be easily derived:

$$\frac{1}{d_{hk0}^2} = \frac{h^2 + k^2}{a^2} \tag{3.6}$$

From this Equation (3.6),

$$d_{hk0} \propto \frac{1}{\sqrt{h^2 + k^2}} \tag{3.7}$$

When the (hk) values are substituted into this equation, the following ratios can be obtained.

$$d_{100} : d_{110} : d_{200} : d_{210} : d_{300} : d_{220} : \cdots$$
$$= 1 : \frac{1}{\sqrt{2}} : \frac{1}{2} : \frac{1}{\sqrt{5}} : \frac{1}{3} : \frac{1}{\sqrt{8}} : \cdots \tag{3.8}$$

Comparing these present ratios with previous two sets of ratios in Equations (3.3) and (3.5), the characteristic reflections having the ratios such as $\frac{1}{\sqrt{2}}$, $\frac{1}{\sqrt{5}}$ and $\frac{1}{\sqrt{8}}$ are observed. This state corresponds to the cases of the S_T and Col_{tet} phases. The characteristic reflections serve as a good guideline for the identification as the S_T and Col_{tet} phases.

As described above, it can be seen that with increasing the temperature the 3D crystal lattices disintegrate stepwise into the corresponding [2D$\oplus$1D] and 1D lattices, and finally into 0D of an isotropic liquid (cf. Fig. 3.7).

3.3.1.4 *Lattice disintegration in monoclinic crystal system*

As far as the author knows, all liquid crystal phases known to date are derived only from the four crystal systems: orthorhombic, hexagonal, tetragonal and monoclinic.

So, finally, we consider the lattice disintegration in the monoclinic crystal system. In this system, we should consider the following three cases. As can be seen in Table 3.3, there are three different tilt directions in the monoclinic system.

(Case 1) The a-axis is inclined in the b-axis direction. In other words, the angle γ is not 90°.

(Case 2) The c-axis is inclined in the a-axis direction. In other words, the angle β is not 90°.

(Case 3) The c-axis is inclined in the b-axis direction. In other words, the angle α is not 90°.

As can be seen from Table 3.3, Case 1 gives a discotic oblique ordered columnar ($\text{Col}_{\text{ob.o}}$) liquid crystal phase. On the other hand, Case 2 and Case 3 give smectic F (S_F) and smectic I (S_I) phases, respectively. Thus, these $\text{Col}_{\text{ob.o}}$, S_F and S_I liquid crystal phases can be derived from the stepwise lattice disintegration of the monoclinic crystal system.

Table 3.3 also summarizes relationship among spacing (d), Miller indices (hkl) and the lattice constants ($a, b, c, \alpha, \beta \text{ and } \gamma$) for the [2D⊕1D] mesophases known to date. As can be seen from this table, each of the [2D⊕1D] **composite-lattice-based liquid crystal phases** is derived from the corresponding specific 3D crystal system [4].

3D hexagonal crystal system → Col_{ho} and S_B
3D orthorhombic crystal system → Col_{ro} and S_E
3D tetragonal crystal system → $\text{Col}_{\text{tet.o}}$ and S_T

From 3D monoclinic crystal system,

when $\gamma \neq 90°$, $\text{Col}_{\text{ob.o}}$
when $\beta \neq 90°$, S_F
when $\alpha \neq 90°$, S_I

Apart from these phases, the other phases S_L, S_G, S_H, S_J and S_K enclosed in parentheses in Table 3.3 are "rigid" [2D⊕1D] liquid crystal phases having only a little bit of three-dimensionality [5] (See the detailed description of (3.1.3.1)).

As described above, in the process of disintegration of a three-dimensional crystal lattice, the corresponding liquid crystal phases are formed between the crystal and the liquid. Table 3.3 summarizes the expressions of the original four crystal systems and the corresponding liquid crystal phases. As can be easily seen also from this table, each of the [2D⊕1D] **composite-lattice-based liquid crystal phases** gives two-dimensional reflection $(hk0)$ lines and one-dimensional reflection (00ℓ) lines, but no three-dimensional reflection $(hk\ell)$ lines. In Table 3.4 is summarized relationship between phase dimensionality and X-ray reflection line dimensionality. For example, a phase having 3D of dimensionality gives all the reflections having 1-dimensional (00ℓ) lines, 2-dimensional $(hk0)$ lines, and 3-dimensional $(hk\ell)$ lines. On the other hand, a phase having [2D⊕1D] of dimensionality gives only the reflections having 1-dimensional (00ℓ) lines and 2-dimensional $(hk0)$ lines. Thus, the dimensionality of any phase corresponds well to the dimensionality of X-ray reflections [4].

3.3.2 *How the dimensionality of X-ray reflection lines appear*

Figure 3.8 schematically shows the appearance of the dimensionality of the X-ray reflection lines of the S_B phase and the Col_{ho} phase. As described in Section 3.3.1.2, the relationship between the spacing d, the Miller index, and the lattice constant is the same for of the S_B phase and the Col_{ho} phase, as expressed in the Equations (3.4.1) and (3.4.2). Here, we will consider how to distinguish the S_B phase from the Col_{ho} phase from the X-ray diffraction pattern, although it can be expressed by the same equations.

As can be seen from Fig. 3.8(1), the liquid crystalline smectic B (S_B) phase gives the two-dimensional (2D) reflection line in the wide angle region, and the one-dimensional (1D) reflection lines in the low angle region. Quite oppositely, as can be seen from Fig. 3.8(2), the columnar liquid crystalline Col_{ho} phase gives the 2D reflection lines in the low angle region, and the 1D reflection line in the wide angle side. This difference is due to the difference in the molecular sizes between rod-like molecules and discotic molecules. As can be seen from the left side of Fig. 3.8, the molecule exhibiting a smectic liquid

(1) The case of S$_B$ phase

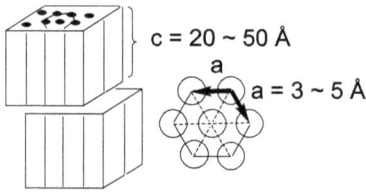

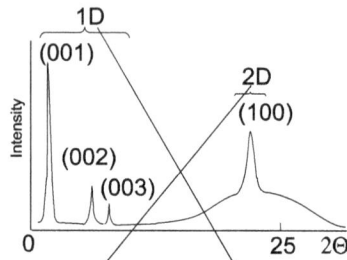

(2) The case of Col$_{ho}$ phase

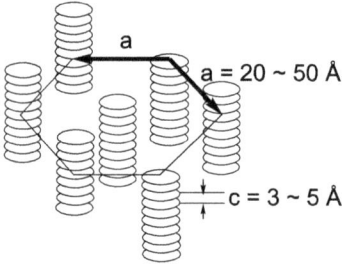

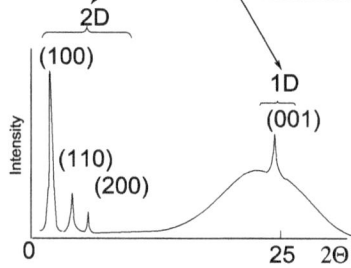

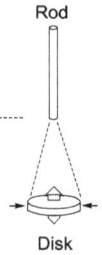

Figure 3.8.   Schematic illustrations of X-ray diffraction patterns of a S$_B$ phase (1) and a Col$_{ho}$ phase (2). The S$_B$ phase gives the one-dimensional (1D) reflection lines in the low angle region and the two-dimensional (2D) reflection line in the wide angle region, whereas the Col$_{ho}$ phase gives the 2D reflection lines in the low angle region and the 1D reflection line in the wide angle region. This is due to the difference in molecular sizes between rod-like (1) and disk-like (2) molecules.

crystal phase is usually rod-like, and its diameter is about 3 to 5 Å, and its length is about 20 to 50 Å. On the other hand, the molecules exhibiting a columnar liquid crystal phase are usually disc-shaped, and their diameter is about 20 to 50 Å, and their thickness is about 3 to 5 Å.

Here from Bragg's condition:

$$2d \sin \theta = \lambda \text{ (when } n = 1) \quad (10) \text{ (cf. Chapter 1 formula (1))}$$

Hence a following Equation (3.11) can be derived:

$$d \propto \frac{1}{\sin \theta} \propto \frac{1}{\theta} \tag{3.11}$$

Table 3.5.   Inverse relationship between d and θ.

d		2θ
Small	$\leftrightarrow$	Big
Big	$\leftrightarrow$	Small

From Bragg's condition, $2d \cdot \sin\theta = n\lambda$.
When $n = 1, 2d \cdot \sin\theta = \lambda$.
$\therefore d\sin\theta = \frac{\lambda}{2} (= \text{constant})$
$\therefore d \propto \frac{1}{\sin\theta} \propto \frac{1}{\theta}$

As is clear from this equation, when the spacing d is large, θ is small; when the spacing d is small, θ is large (Table 3.5). Therefore, a reflection line corresponding to the dimension of 3 to 5 Å appears in the wide angle side where θ is large, and a corresponding reflection line in the dimension of 20 to 50 Å appears in the low angle side where θ is small. In the case of a smectic liquid crystal phase, a one-dimensional lamellar (layered) structure with a size of 20 to 50 Å appears in the low angle region, and a two-dimensional packing of rod-like molecules corresponding to a size of 3 to 5 Å appears in the wide angle region. On the other hand, in the case of the columnar liquid crystal phase, the two-dimensional packing of columns with dimensions of 20 to 50 Å appears in the low angle region, and the one-dimensional stacking of the disc-like molecules corresponding to the dimension of 3 to 5 Å appears in the low angle region.

As described above, the two-dimensionality of the columnar liquid crystal phase appears in the low angle region, whereas the two-dimensionality of the smectic (lamellar) liquid crystal phase appears quite oppositely in the wide angle region [4]. Therefore, a pair of the smectic liquid crystal phase and columnar phase, i.e., (S_B and Col_{ho}), (S_E and Col_{ro}), (S_T [6] and $Col_{tet.o}$), etc. can be expressed by the same formula, but they are distinguishable from the appearance of the dimensionality in the X-ray diffraction patterns.

Considering in this way, the smectic phase of the rod-like molecules and the columnar phase of the disc-like molecules can be considered uniformly. For example, the liquid crystal phases S_E and

Col$_{ro}$ can be represented by the same two Equations (3.2.1) and (3.2.2), and the dimensionality is the same [2D⊕1D] dimension. If the two-dimensional lattices have the same symmetry of P2$_1$, they are the completely same liquid crystal phase. Only the dimensions of the molecules differ from rod to disk. Therefore, there is also a claim that the S$_E$ phase is a crystal E (crystal phase), but since the Col$_{ro}$ phase is recognized as a liquid crystal phase, the S$_E$ phase should be also a liquid crystal phase. Thus, both the S$_E$ phase and the Col$_{ro}$ phase have [2D⊕1D] dimensionality expressed by the same equations, and have no three-dimensionality characteristic to the crystal phase This proves that the S$_E$ phase is not a crystalline phase but a liquid crystal phase. Similarly, the liquid crystal phases S$_B$ and Col$_{ho}$ phases can be represented by the same two Equations (3.4.1) and (3.4.2), and the dimensionality is the same [2D⊕1D] dimension. Moreover, each of these phases has the same six-fold rotational symmetry in their two-dimensional lattices. Therefore, they are exactly the same liquid crystal phase. Only the dimensions of the molecules differ from rod to disk. Therefore, there is also a claim that the S$_B$ phase is a crystal B (crystal phase), but since the Col$_{ho}$ phase is recognized as a liquid crystal phase, the S$_B$ phase should be also a liquid crystal phase. Thus, both the S$_B$ phase and the Col$_{ho}$ phase have [2D⊕1D] dimensionality expressed by the same equations, and have no three-dimensionality characteristic to a crystalline phase. This proves that the S$_B$ phase is not a crystalline phase but a liquid crystal phase. It should be noted that the [2D⊕1D] dimension is not three-dimensional (see Section 3.3.5).

3.3.3 *Liquid crystal phase is subspace*

Considering in the same way above-mentioned, the dimensionality of the molecular assemblies is stepwise disintegrated from 3D of crystal to [2D⊕1D], 2D, [1D⊕1D], 1D, etc, of liquid crystal and finally 0D of liquid. Thus, it can be seen that liquid crystal phases appear as various dimensional assemblies in the process of stepwise disintegration of the three-dimensional crystal lattice.

"**In terms of the language of linear algebra, the crystal phase is a three-dimensional space, and the liquid crystal phase is its subspace.**" Here, the notation "2D⊕1D" or "1D⊕1D" means a direct sum of subspaces. The [2D⊕1D] dimensional space is not a three-dimensional space but a direct sum of a two-dimensional subspace and a one-dimensional subspace. Similarly, the [1D⊕1D]-dimensional space is not a two-dimensional space but a direct sum of a one-dimensional subspace and a one-dimensional subspace. Therefore, the liquid crystal phase is a subspace. With this concept of "stepwise dimensionality disintegration" matter can be regarded as a dimensional assembly from the viewpoint of linear algebra. When structural analysis is carried out, there may be 3D, 2D, 1D, [2D⊕1D], [1D⊕1D], and [1D⊕1D⊕1D] dimension set states (phases) can be expected.

As can be seen from Table 3.6,

2D liquid crystal phases are Col_{hd}, Col_{rd}, $Col_{tet.d}$, etc.
1D liquid crystal phases are S_A, S_C, $N_C (= Col_N)$, etc.
[2D⊕1D] liquid crystal phases are S_B, S_E, S_T [6], Col_{ho}, Col_{ro}, and $Col_{tet.o}$, etc.
[1D⊕1D] liquid crystal phases are $D_{L2} (= D_{LC} = Col_L)$, etc.
[1D⊕1D⊕1D] liquid crystal phases: two examples have been reported so far [4], although this liquid crystal phase has not been named.

Among these dimensional assemblies, the liquid crystal phases having [2D⊕1D], [1D⊕1D], and [1D⊕1D⊕1D] are **composite-lattice-based liquid crystal phases** as already explained in Section 3.3.3. In Table 3.4, the lattice types are denoted as S and C for the single-lattice-based phase and the composite-lattice-based phase, respectively.

3.3.4 *Matter is not three states but multiple states*

In Section 3.1 and Table 3.1, the idea for the conventional liquid crystal phase was described. There, the "**long-distance 3D order**

Table 3.6. Original crystal systems and dimensionalities of liquid crystalline phase resulted by stepwise lattice disintegration.

Rod-like LC mesophase	Tilt	Original crystal system	Order in the layer	Phase Dimensionality$	Discotic LC mesophase	Tilt	Original crystal system	Order in the face	Phase Dimensionality$
N				0D	N_D				0D
S_A	⊥		Non (1D-layer)	1D					
S_C	∠		Non (1D-layer)	1D	D_{L1}	∠		Non (1D-layer)	1D
S_D	—	3D-cubic	—	3D	Col_{hd}		3D-hexagonal	2D-hexagonal	2D
S_B	⊥	3D-hexagonal	2D-hexagonal	[2D⊕1D]	$Col_{tet.d}$		3D-tetragonal	2D-tetragonal	2D
S_T	⊥	3D-tetragonal	2D-tetragonal	[2D⊕1D]	Col_{rd}		3D-orthorohmbic	Rectangular	2D
S_E	⊥	3D-orthorohmbic	Rectangular	[2D⊕1D]	$Col_{ob.d}$	∠γ	3D-monoclinic	2D-pallarelogram	2D
S_F	∠β	3D-monoclinic.	Rectangular	[2D⊕1D]	Col_{ho}	⊥	3D-hexagonal	2D-hexagonal	[2D⊕1D]
S_I	∠α	3D-monoclinic.	Rectangular	[2D⊕1D]	$Col_{tet.o}$	⊥	3D-tetragonal	2D-tetragonal	[2D⊕1D]
S_L	⊥	3D-hexagonal	2D-hexagonal	3D <[2D⊕1D]	Col_{ro}	⊥	3D-orthorohmbic	Rectangular	[2D⊕1D]
S_G	∠β	3D-monoclinic.	Rectangular	3D <[2D⊕1D]	$D_{L,rec}(2_1 2_1)$	⊥	3D-orthorohmbic	Rectangular	[2D⊕1D]
S_H	∠β	3D-monoclinic	Rectangular	3D <[2D⊕1D]	$D_{L,rec}(12_1)$	⊥	3D-orthorohmbic	Rectangular	[2D⊕1D]
S_J (G')	∠α	3D-monoclinic.	Rectangular	3D <[2D⊕1D]	$Col_{ob.o}$	∠γ	3D-monoclinic	2D-pallarelogram	[2D⊕1D]
S_K (H')	∠α	3D-monoclinic	Rectangular	3D <[2D⊕1D]					
					D_{L2} (= Col_L)	⊥			[1D⊕1D]
					No name	⊥			[1D⊕1D]

$: Phase dimensionality can be determined by the X-ray reflection lines; 1D: (00l); 2D: (hk0); 3D: (hkl). α: angle α is not 90°; β: angle β is not 90°; γ: angle γ is not 90°.

of gravity center position" exists in the 3D crystal, but neither the liquid nor the liquid crystal has this order, so that liquid crystal is zero-dimensional (0D) marked with × in this table. However, as can be seen from Table 3.6, the liquid crystal phase having 0D is only the nematic liquid crystal phase, whereas many other liquid crystal phases have at least 2D or 1D property. Therefore, it can be seen that the conventional concept for liquid crystal applies only to the nematic liquid crystal phase and does not cover various liquid crystal phases. Accordingly, we have decided to consider "long-distance order of gravity center position" not only as 3D but also as 2D and 1D. The dimensionality of the lattice of each phase is expressed as 3D, $[2D \oplus 1D]$, $[1D \oplus 1D \oplus 1D]$, 2D, $[1D \oplus 1D]$, 1D and 0D.

The conventional idea that the dimensionality changes directly from 3D crystal to 0D liquid in the phase transition of matter is obviously wrong, as you can see from Fig. 3.7. **Matter is not three states of (crystal → liquid → vapor) but matter is multi-states**. The dimensionality changes in a stepwise manner like as $3D \rightarrow [2D \oplus 1D] \rightarrow [1D \oplus 1D] \rightarrow 1D \rightarrow 0D$. Thus, the 3D crystal lattice stepwise disintegrates to 0D. The author's idea of "**stepwise dimensionality disintegration**" has never been clearly written in the textbooks of liquid crystals until now, to our best knowledge. The present classification of the liquid crystal phases according to the dimensionality can be applied to both rod-like liquid crystals and discotic liquid crystals, both thermotropic liquid crystals and lyotropic liquid crystals, i.e., to any kind of liquid crystals. Furthermore, since it well corresponds to X-ray reflection lines, it is extremely useful. From this new classification, we can also readily understand that "**the dimensionality of the liquid crystal phases has a hierarchy**."

3.4 Extinction Rule of a Two-dimensional Lattice

As mentioned above, the stepwise disintegration of the crystal phase produces the liquid crystal phase. Therefore, when X-ray structural analysis of a liquid crystal phase is actually performed, a two-dimensional lattice or a one-dimensional lattice is handled.

The analysis of a one-dimensional lattice is very simple, since only Equation (3.2.2) is used. On the other hand, the analyses of two-dimensional lattices are much more difficult, because we must additionally use the extinction rules of the two-dimensional lattices. Therefore, next we consider the extinction rules of the two-dimensional lattices. Up to date, there has never been such a book that describes how to calculate the extinction rules of two-dimensional lattices, as far as the author knows. The author strongly believes that the following explanations will be very useful for the readers.

From Equation (1.22) in Chapter 1
Crystal structure factor:

$$F(hkl) = \sum_i^N f_i \cdot \exp\{2\pi i(hx_i + ky_i + lz_i)\} \qquad (3.12)$$

If this equation is reduced into two dimensionality.
Liquid crystal structure factor:

$$F(hk) = \sum_i^N f_i \cdot \exp\{2\pi i(hx_i + ky_i)\} \qquad (3.13)$$

Here, this two-dimensional structure factor is called a **liquid crystal structure factor**. This is served to calculate the extinction rules of two-dimensional lattices having various symmetries.

3.4.1 *Extinction Rule of Two-dimensional tetragonal lattice*

The extinction rule of a two-dimensional tetragonal lattice is calculated by the same method as that for the extinction rule of NaCl in a face-centered cubic lattice in Chapter 1 (1.12.2).

First, as shown in Fig. 3.9, it is assumed that four equivalent molecules A form a two-dimensional tetragonal lattice. At this time, the Cartesian coordinates of each of the molecules A become like the leftmost column of Table 3.7. In addition, the number of each

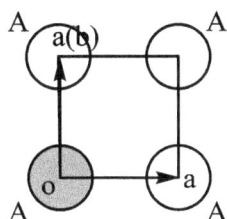

Figure 3.9. Col$_{tet}$ lattice.

Table 3.7. Derivation of the extinction rule for two-dimensional tetragonal lattice.

Coordinate of A	n in nf_A	$\exp\{2\pi i(hx_i + ky_i)\}$
(0 0)	$\frac{1}{4}$	$\exp\{0\} = 1$
(1 0)	$\frac{1}{4}$	$\exp\{2\pi i(h)\} = 1$
(1 1)	$\frac{1}{4}$	$\exp\{2\pi i(h + k)\} = 1$
(0 1)	$\frac{1}{4}$	$\exp\{2\pi i(k)\} = 1$

Therefore, the sum total F(hk) is

$$F(hk) == f_A \left[\frac{1}{4} \times 1 + \frac{1}{4} \times 1 + \frac{1}{4} \times 1 + \frac{1}{4} \times 1\right] = f_A$$

∴ No extinction rules.

molecule contained in one lattice is $1/4$, so the number is written in the center column of this table. In the rightmost column of this table, is entered the calculation result of the part of $\exp\{2\pi i(hx_i + ky_i)\}$ in the liquid crystal structure factor.

From Table 3.7, the sum $F(hk)$ of liquid crystal structure factors can be calculated as follows:

$$F(hk) = f_A \left[\frac{1}{4} \times 1 + \frac{1}{4} \times 1 + \frac{1}{4} \times 1 + \frac{1}{4} \times 1\right] = f_A \qquad (3.14)$$

Therefore, there is no extinction rule for this two-dimensional tetragonal lattice.

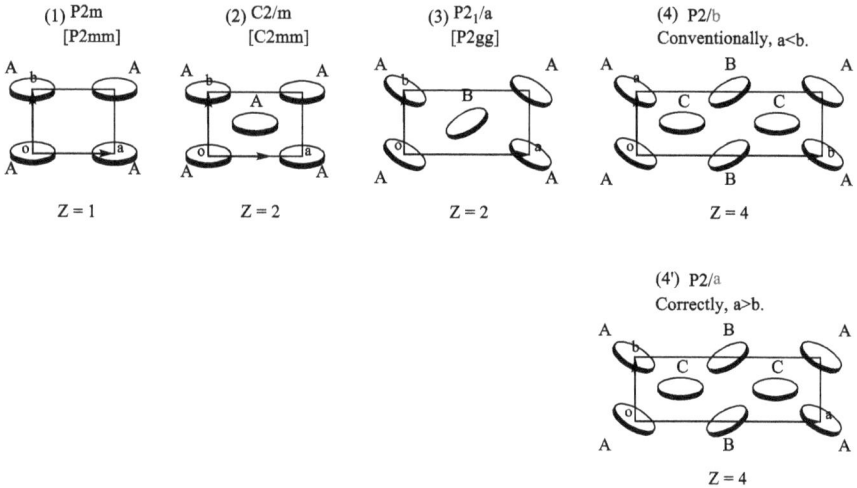

Figure 3.10. Col_r lattices.

3.4.2   *Extinction rule of a two-dimensional rectangular lattice*

Figure 3.10 illustrates four types of rectangular lattices that have been discovered so far. These symmetries are (1) $P2m$ [another notation: $p2mm$], (2) $C2/m$ [$C2mm$], (3) $P2_1/a$ [$p2gg$], (4) $P2/b$ (conventionally, $a < b$), and (4') $P2/a$ (correctly, $a > b$).

3.4.2.1   *In the case of P2m [p2mm] symmetry*

As you can see immediately from Fig. 3.10(1), this rectangular lattice having $P2m$ [$p2mm$] symmetry is equivalent to the two-dimensional tetragonal lattice in Fig. 3.9. Therefore, when the liquid crystal structure factor is calculated for this case of $P2m$ symmetric lattice, we have exactly the same result as the two-dimensional tetragonal lattice. Hence, there is no extinction rule for this $P2m$ symmetric lattice.

3.4.2.2   *In the case of C2/m [C2mm] symmetry*

As shown in Fig. 3.10(2), five equivalent molecules A form a two-dimensional C bottom-centered lattice. At this time, the Cartesian

Table 3.8. Derivation of the extinction rule for symmetry of $C2/m$.

Coordinate of A	n in nf_A	$\exp\{2\pi i(hx_i + ky_i)\}$
(0 0)	$\dfrac{1}{4}$	$\exp\{0\} = 1$
(1 0)	$\dfrac{1}{4}$	$\exp\{2\pi i(h)\} = 1$
(1 1)	$\dfrac{1}{4}$	$\exp\{2\pi i(h + k)\} = 1$
(0 1)	$\dfrac{1}{4}$	$\exp\{2\pi i(k)\} = 1$
$\left(\dfrac{1}{2}\dfrac{1}{2}\right)$	1	$\exp\left\{2\pi i\left(\dfrac{1}{2}h + \dfrac{1}{2}k\right)\right\} = (-1)^{h+k}$

Therefore, the sum total F(hk) is

$$F(hk) = f_A\left[\frac{1}{4} \times 1 + \frac{1}{4} \times 1 + \frac{1}{4} \times 1 + \frac{1}{4} \times 1 + 1(-1)^{h+k}\right]$$

$$= f_A[1 + (-1)^{h+k}]$$

$\therefore$ When $h + k = 2n$, $F(hk) = 2f$
When $h + k = 2n + 1$, $F(hk) = 0$ (extinction)

coordinates of each of the molecules A become like the leftmost column of Table 3.8. In addition, since the number of molecules contained in one lattice is 1 for the molecule at the center and 1/4 each for the four molecules at the periphery. These numbers are written in the center column of this table. In the rightmost column of this table, is entered the calculation result of the part of $\exp\{2\pi i(hx_i + ky_i)\}$ in the liquid crystal structure factor.

Therefore, the sum $F(hk)$ of liquid crystal structure factors is calculated as follows:

$$F(hk) == f_A\left[\frac{1}{4} \times 1 + \frac{1}{4} \times 1 + \frac{1}{4} \times 1 + \frac{1}{4} \times 1 + 1 \times (-1)^{h+k}\right]$$

$$= f_A\left[1 + (-1)^{h+k}\right] \tag{3.15}$$

From this factor, we can derive the following rules:

When $h + k = 2n$, $F(hk) = 2f$

When $h + k = 2n + 1$, $F(hk) = 0$ (extinction) $\tag{3.16}$

Equation (3.16) is the extinction rule in the case of $C2/m$ [$C2mm$] symmetry.

3.4.2.3 *In the case of $P2_1/a$ [p2gg] symmetry*

As shown in Fig. 3.10(3), four equivalent molecules A in the periphery and one non-equivalent molecule B in the center form a rectangular two-dimensional lattice. Since the X-ray reflectivities of the molecule A and the molecule B are different, their reflectivities are denoted as f_A and f_B respectively. At this time, the liquid crystal structure factor can be calculated, as shown in Table 3.9. The sum $F(hk)$ of the liquid crystal structure factor is

$$F(hk) = f_A \left(\frac{1}{4} \times 1 + \frac{1}{4} \times 1 + \frac{1}{4} \times 1 + \frac{1}{4} \times 1 \right) + f_B[1 \times (-1)^{h+k}]$$

$$= f_A + f_B(-1)^{h+k} \tag{3.17}$$

Table 3.9.   Derivation of the extinction rule for symmetry of $P2/a$.

Coordinate of A	n in nf_A	$\exp\{2\pi i(hx_i + ky_i)\}$
(0 0)	$\dfrac{1}{4}$	$\exp\{0\} = 1$
(1 0)	$\dfrac{1}{4}$	$\exp\{2\pi i(h)\} = 1$
(1 1)	$\dfrac{1}{4}$	$\exp\{2\pi i(h+k)\} = 1$
(0 1)	$\dfrac{1}{4}$	$\exp\{2\pi i(k)\} = 1$
Coordinate of B	n in nf_B	$\exp\{2\pi i(hx_i + ky_i)\}$
$\left(\dfrac{1}{2}\dfrac{1}{2}\right)$	1	$\exp\left\{2\pi i\left(\dfrac{1}{2}h + \dfrac{1}{2}k\right)\right\} = (-1)^{h+k}$

Therefore, the sum total $F(hk)$ is

$$F(hk) = f_A \left(\frac{1}{4} \times 1 + \frac{1}{4} \times 1 + \frac{1}{4} \times 1 + \frac{1}{4} \times 1 \right) + f_B[1 \times (-1)^{h+k}]$$

$$= f_A + f_B(-1)^{h+k}$$

$\therefore$ When $h + k = 2n$, $F(hk) = f_A + f_B$ (strong reflection).
When $h + k = 2n + 1$, $F(hk) = f_A - f_B$ (weak reflection).

(3) P2₁/a
[P2gg]

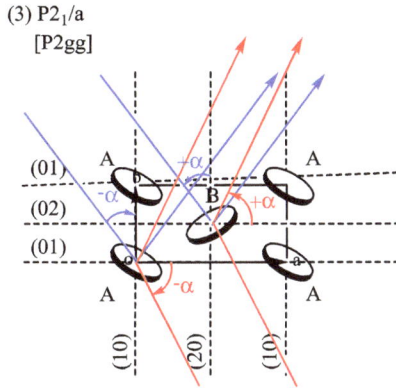

When h0: h = 2n + 1, F(hk) = 0
When 0k: k = 2n + 1, F(hk) = 0

Figure 3.11. Col$_r$(P2₁/a).

From this factor, we can derive the following rules:

When $h + k = 2n$, $\quad F(hk) = f_A + f_B$(strong intensity).

When $h + k = 2n + 1$, $\quad F(hk) = f_A - f_B$(weak intensity).

$$(3.18)$$

Looking at these results, there seems to be no condition being $F(hk) = 0$. However, when considering the symmetry, there is a condition being $f_A = f_B$. That is the case for $h = 0$ or $k = 0$.

As shown in Fig. 3.11, when considering the case of $h = 0$, the reflective surface is inclined at $-\alpha$ for molecule A and $+\alpha$ for molecule B with respect to the b-axis. In this case, the reflectivity is $f_A = f_B$. For $k = 0$, $f_A = f_B$ for the same reason.

Therefore,

$$h0 : \text{when } h = 2n + 1, \ F(hk) = 0$$

$$0k : \text{when } k = 2n + 1, \ F(hk) = 0 \ (\because (3.18)) \qquad (3.19)$$

Equations (3.19) are the extinction rule in the case of $P2_1/a$ [$p2gg$] symmetry.

Therefore, in the case of this $P2_1/a$ symmetry, the reflection from the (3.21) plane is weak but observed ($\because$ (3.18)). On the other hand, in the case of the above-mentioned $C2/m$ symmetry, no reflection from the (3.21) plane is observed ($\because$ (3.16)).

3.4.2.4 *In the case of P2/b (conventionally, a < b) symmetry*

As shown in Fig. 3.10(4), three kinds of molecules A, B, and C with different reflectivities form a two-dimensional rectangular lattice. The X-ray reflectivity of the molecules A, B and C is denoted as f_A, f_B and f_C, respectively. At this time, the liquid crystal structure factor can be calculated, as shown in Table 3.10. The sum $F(hk)$ of the liquid crystal structure factor is

$$F(hk) = f_A + f_B \cdot (-1)^k + f_C \cdot (-1)^h \left\{ i \sin\left(\frac{\pi}{2}k\right) + i \sin\left(\frac{3\pi}{2}k\right) \right\}$$
(3.20)

As shown in Fig. 3.12(4), when $h = 0$, $f_A = f_B = f$. Therefore Equation (3.20) can be expressed as

$$F(hk) = f\{1 + (-1)^k\} + f_C \cdot (-1)^h \left\{ i \sin\left(\frac{\pi}{2}k\right) + i \sin\left(\frac{3\pi}{2}k\right) \right\}$$
(3.21)

Accordingly,

When $k = 2n + 1$, $F(hk) = 0$ (extinction).

Therefore, at this time, it disappears.

Also, as depicted in Fig. 3.12(4), even when $k = 0$, $f_A = f_B = f$

Therefore Equation (3.20) can be expressed as

$$F(hk) = f\left\{ 1 + (-1)^k + f_C \cdot (-1)^h i \sin\left(\frac{\pi}{2}k\right) + i \sin\left(\frac{3\pi}{2}k\right) \right\} = 2f$$
(3.22)

Accordingly

When $h = 2n + 1$, $F(hk) \neq 0$.

Therefore, at this time, it appears.

Thus, the extinction rule in the case of $P2/b$ symmetry can be derived as

$$(0k) : k = 2n + 1, \ F(hk) = 0 \text{ (extinction)}. \qquad (3.23)$$

Table 3.10. Derivation of the extinction rule for symmetry of P2/b.

Coordinate of A	n in nf_A	$\exp\{2pii(hx_i + ky_i)\}$
(0 0)	$\frac{1}{4}$	$\exp\{0\} = 1$
(1 0)	$\frac{1}{4}$	$\exp\{2\pi i(h)\} = 1$
(1 1)	$\frac{1}{4}$	$\exp\{2\pi i(h + k)\} = 1$
(0 1)	$\frac{1}{4}$	$\exp\{2\pi i(k)\} = 1$
Coordinate of B	n in nf_B	$\exp\{2\pi i(hx_i + ky_i)\}$
$\begin{pmatrix} 0\frac{1}{2} \end{pmatrix}$	$\frac{1}{2}$	$\exp\left\{2\pi i\left(\frac{1}{2}k\right)\right\} = (-1)^k$
$(1\frac{1}{2})$	$\frac{1}{2}$	$\exp\left\{2\pi i\left(h + \frac{1}{2}k\right)\right\} = (-1)^k$
Coordinate of C	n in nf_C	$\exp\{2\pi i(hx_i + ky_i)\}$
$\begin{pmatrix} \frac{1}{2}\frac{1}{4} \end{pmatrix}$	1	$\exp\left\{2\pi i\left(\frac{1}{2}h + \frac{1}{4}k\right)\right\} = (-1)^h \cdot i\sin\left(\frac{\pi}{2}k\right)$
$\begin{pmatrix} \frac{1}{2}\frac{3}{4} \end{pmatrix}$	1	$\exp\left\{2\pi i\left(\frac{1}{2}h + \frac{3}{4}k\right)\right\} = (-1)^h \cdot i\sin\left(\frac{3\pi}{2}k\right)$

Therefore, the sum total $F(hk)$ is

$$F_{(hk)} = f_A + f_B \cdot (-1)^k + f_C \cdot (-1)^h \{\sin\left(\frac{\pi}{2}k\right) + i\sin\left(\frac{3\pi}{2}k\right)\}$$

From the reason illustrated in Fig. 3.19, when $h = 0$, $f_A = f_B = f$.

$$f_{(hk)} = f\{1 + (-1)^k\} + f_C \cdot (-1)^h \left\{i\sin\left(\frac{\pi}{2}k\right) + i\sin\left(\frac{3\pi}{2}k\right)\right\}$$

When $k = 2n + 1$, $F(hk) = 0$ (extinction).
From the reason illustrated in Fig. 3.19, when $k = 0$, $f_A = f_B = f$.

$$F(hk) = f\{1 + (-1)^k\} + f_C \cdot (-1)^h \left\{i\sin\left(\frac{\pi}{2}k\right) + i\sin\left(\frac{3\pi}{2}k\right)\right\} = 2f$$

When $h = 2n + 1$, $F(hk) \neq 0$ (appearance).
$\therefore$ When $(0k) : k = 2n + 1$, $F(hk) = 0$ (extinction).

3.4.2.5 *In the case of P2/a (correctly, a > b) symmetry*

When the $P2/a$ symmetric rectangular lattice was first reported [11], the figure being $a < b$ (Fig. 3.10(4)), was mistakenly drawn in the

(4) P2/b

Conventionally, a<b.

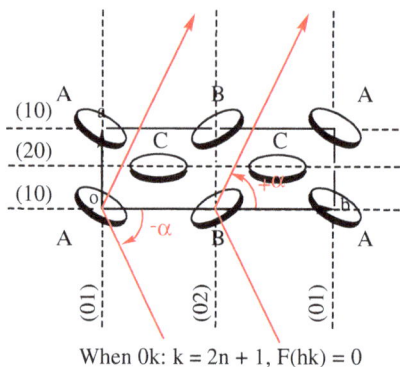

(4') P2/a

Correctlly, a>b.

When 0k: k = 2n + 1, F(hk) = 0

When h0: h = 2n + 1, F(hk) = 0

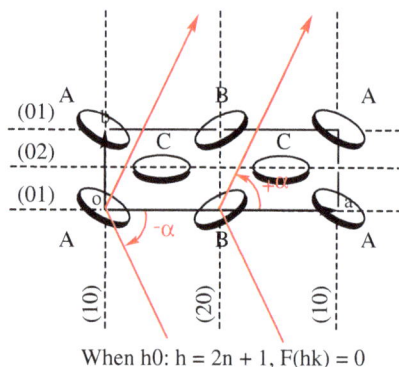

Figure 3.12.   P2/b and P2/a.

paper. The lattice drawn there has a symmetry of P2/b, although it was denoted as P2/a. Therefore, for many years, the extinction rule for this symmetry has been Equation (3.23). However, according to the X-ray structural analysis textbook [7], there is a rule that when writing a two-dimensional lattice, it must be $a > b$. Therefore, in this book, we correctly illustrate a $P2/a$-symmetric lattice having $a > b$ in Fig. 3.10(4′), and extinction rule of this symmetry is calculated in the similar manner explained in Section 3.4.2.4.

The extinction rule for this $P2/a$ symmetry is obtained as

$$(h0): \ h = 2n + 1, \ F(hk) = 0 \text{ (extinction)}. \qquad (3.24)$$

Readers are required from now on, to use the $P2/a$ symmetric lattice of Fig. 3.12(4′) and the extinction rule of (3.24) correctly.

As described above, the extinction rules of four types of two-dimensional rectangular lattices are derived by using their liquid crystal structure factors. In Fig. 3.13 are summarized these extinction rules and the symmetries of four rectangular lattices know to date.

(1) P2m
   [P2mm]

(2) C2/m
   [C2mm]

(3) P2$_1$/a
   [P2gg]

(4) P2/b
Conventionally, a<b.

Z = 1

Z = 2

Z = 2

Z = 4

All appear.
No extinction rules.

hk: h + k = 2n + 1

h0: h = 2n + 1
0k: k = 2n + 1

0k: k = 2n + 1

(4') P2/a
Correctlly, a>b.

Z = 4
h0: h = 2n + 1

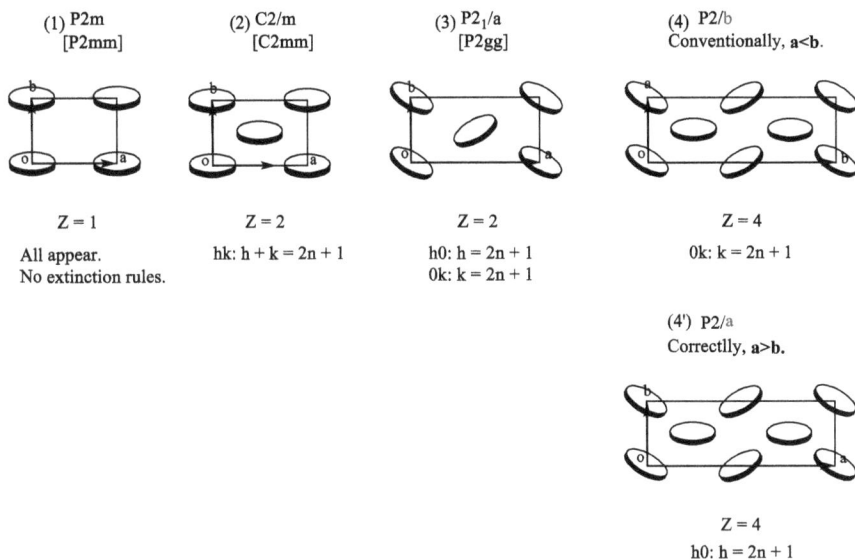

Figure 3.13. Summary of the extinction rules and the symmetries of four rectangular lattices.

3.5 X-ray Structural Analysis of Liquid Crystal Phases

3.5.1 *Golden rule for liquid crystal structure analysis*

Next, when actually carrying out the structure analysis of liquid crystal phases, we should presume from the X-ray diffraction data what two-dimensional lattices or one-dimensional lattices appear. Hereupon, most people get lost the way. It is because that we have never had any guideline to overview the whole.

However, it will be possible to overview the whole analysis without hesitation from a new guideline of "golden rule for liquid crystal structure analysis" that the author has taught to the students and graduate students in Shinshu University. This golden rule will be very helpful for all the readers. The golden rule is as follows:

① If the liquid crystal phase has a one-dimensional layered (lamella) structure, from Equation (3.3), the ratio of spacing d will be given as follows:

$$d_1 : d_2 : d_3 : d_4 : d_5 : \cdots = 1 : \frac{1}{2} : \frac{1}{3} : \frac{1}{4} : \frac{1}{5} : \cdots$$

② If the liquid crystal phase has a two-dimensional hexagonal structure, from Equation (3.5), the ratio of spacing d will be given as follows:

$$d_1 : d_2 : d_3 : d_4 : d_5 : d_6 : \cdots = 1 : \frac{1}{\sqrt{3}} : \frac{1}{2} : \frac{1}{\sqrt{7}} : \frac{1}{3} : \frac{1}{\sqrt{12}} : \cdots$$

③ If the liquid crystal phase has a two-dimensional tetragonal structure, from Equation (3.8), the ratio of spacing d will be given as follows:

$$d_1 : d_2 : d_3 : d_4 : d_5 : d_6 : \cdots = 1 : \frac{1}{\sqrt{2}} : \frac{1}{2} : \frac{1}{\sqrt{5}} : \frac{1}{3} : \frac{1}{\sqrt{8}} : \cdots$$

④ If the liquid crystal phase has a two-dimensional rectangular structure, it cannot be represented by a simple ratio such as the above-mentioned articles from ① to ③. However, we sometimes notice that the ratio shows two series of layer structures (the nature of Article ①). This is the feature of the rectangular structure.

When this phase is likely to be a rectangular phase, the two strong reflections in the lowest angle region are assumed as the reflections of (20) and (11) (or (11) and (20)). By substituting the spacing values of d_{20} and d_{11}, the lattice constants, a and b, are calculated from the following equation:

$$\frac{1}{d_{hk0}^2} = \frac{h^2}{a^2} + \frac{k^2}{b^2}$$

You draw a reciprocal lattice by using these obtained lattice constants, a and b, and carry out the analysis. Then, you use the extinction rules (Fig. 3.13) to determine the symmetry of the lattice.

If you know the above Articles ① to ④, you can perform X-ray structural analysis of almost all liquid crystal phases in rod-like, discotic, thermotropics, lyotropic, low molecular, and polymer liquid crystals.

3.5.2 *Analysis of liquid crystal phase by "Reciprocal Lattice Method"*

First of all, the ratio of the spacing d is calculated from the X-ray diffraction data of the liquid crystal phase. Next, the ratio is examined sequentially from the first article of "golden rule for liquid crystal structure analysis," in order to presume the dimensionality (1D or 2D) of the liquid crystal lattice.

When this liquid crystal phase shows only one-dimensionality corresponding to the first article in the golden rule, it can be easily identified as a lamellar liquid crystal (S_A, S_C, etc.). However, it is basically impossible to distinguish between S_A and S_C only from X-ray structural analysis, so that the final identification is done by the microscopic observations of their textures.

On the other hand, when there is two-dimensionality corresponding to Articles ② to ④, it is not recommended to identify only from the ratio. Although such **Ratio Method** has been used all over the world, unfortunately there have been many mistakes that the liquid crystal phase was incorrectly identified to a wrong phase, and that even if it was identified to a correct phase, the X-ray reflection lines were incorrectly indexed. Therefore, in the case of two-dimensional lattices, it is safe to draw a reciprocal lattice and to identify the phase by using the following novel method of **"Reciprocal Lattice Method."**

The 3D reciprocal lattices and the real lattices were already explained in detail in Chapter 1, Section 1.11 (see Fig. 1.22 and Table 1.3). The present relationship between the 2D real lattice and the reciprocal lattice is summarized in Table 3.11.

There are only four kinds of the two-dimensional lattices in liquid crystals (see Table 3.6): two-dimensional hexagonal lattice (Col_h, S_B, etc.), two-dimensional tetragonal (square) lattice (Col_{tet}, S_T), two-dimensional rectangular lattice (Col_r, S_E, etc.) and two-dimensional oblique lattice (Col_{ob}, S_F, S_I, etc.). However, we rarely encounter the two-dimensional oblique lattice. Therefore, it is almost enough to know really the rest three 2D lattices. Hence, hereafter, the analysis examples using **Reciprocal Lattice Method** will be described for eight discotic phases and three smectic phases listed in Table 3.12.

Table 3.11. Relationship between a real lattice having one-dimensionality or two-dimensionality and the corresponding reciprocal lattice.

1D-lameller lattice: $\dfrac{1}{d_{00l^2}} = \dfrac{l^2}{c^2}$

$$c = \frac{1}{c^*} \quad c^* = \frac{1}{c}$$

2D-hexagonal lattice: $\dfrac{1}{d_{hk0^2}} = \dfrac{4}{3}\left(\dfrac{h^2 + hk + k^2}{a^2}\right)$

$$a = \frac{2}{a^*\sqrt{3}} \quad a^* = \frac{2}{a\sqrt{3}}$$

$$\gamma = 120° \quad \gamma^* = 180° - \gamma = 60°$$

2D-tetragonal lattice: $\dfrac{1}{d_{hk0^2}} = \dfrac{h^2 + k^2}{a^2}$

$$a = \frac{1}{a^*} \quad a^* = \frac{1}{a}$$

$$\gamma = 90° \quad \gamma^* = 180° - \gamma = 90°$$

2D-rectangular lattice: $\dfrac{1}{d_{hk0^2}} = \dfrac{h^2}{a^2} + \dfrac{k^2}{b^2}$

$$a = \frac{1}{a^*} b = \frac{1}{b^*} \quad a^* = \frac{1}{a} \quad b^* = \frac{1}{b}$$

$$\gamma = 90° \quad \gamma^* = 180° - \gamma = 90°$$

2D-oblique lattice: $\dfrac{1}{d_{hk0^2}} = \dfrac{1}{(\sin\gamma)^2}\left(\dfrac{h^2}{a^2} + \dfrac{k^2}{b^2} - \dfrac{2hk\cos\gamma}{ab}\right)$

$$a = \frac{1}{a^*\sin\gamma^*} b = \frac{1}{b^*} \quad a^* = \frac{1}{a\sin\gamma} \quad b^* = \frac{1}{b}$$

$$\gamma = 180° - \gamma^* \quad \gamma^* = 180° - \gamma$$

d: spacing; h, k, l: Miller index; a, b, c and γ: lattice parameter; a^*, b^*, c^* and γ^*: reciprocal lattice parameters.

3.5.3 *Liquid crystal phase having a two-dimensional rectangular lattice (Col$_r$)*

3.5.3.1 *Col$_{rd}$(C2/m) phase*

Figure 3.14 shows the X-ray diffraction pattern at 150°C of the discotic compound $C_{12}PzCu$ [8] which are substituted by eight long

Table 3.12.   X-ray data of representative mesophases.

Mesophases No.	M1	M2	M3	M4	M5
d_{obs}(Å)	24.7	37.4	38.8	No. 1 28.9	32.0
	14.2	21.6	28.4	No. 2 26.3	28.3
	12.3	18.7	20.0	No. 3 16.4	18.6
	9.24	14.2	14.4	No. 4 14.7	12.6
	ca.4.8	10.7	13.1	No. 5 13.2	11.2
	3.36	9.26	ca. 4.2	No. 6 ca.4.7	ca. 4.6
		8.51		No. 7 ca. 3.4	
		ca. 4.6			
Ref. No.	1	10	7	5	36
Material Abbr.	Ni(12,$\bar{1}$)	$[(c_{14}O)_2PhO]_8PcCu$	$[(C_{18}OPh)_8Pc]_2Lu$	$C_{12}PzCu$	C_{10}-Cu
T_{obs}(°C)	170	98	rt	150	100
Mol.weight	1330.61	4710.87	6713.33	1930.58	3685.15
Identified phase	Col_{ho}	Col_{hd}	$Col_{tet.d}$	$Col_{rd}(C2/m)$	$Col_{rd}(P2_1/a)$

	M6	M7	M8	M9	M10	M11
	24.96	20.4	25.7	32.8	25.2	32.9
	17.47	18.6	18.4	16.7	13.1	16.1
	16.00	13.6	16.1	11.2	4.54	10.7
	9.76	10.2	12.6	ca.4.9	4.06	7.97
	8.98	8.56	11.2		3.26	6.35
	ca.5.6	6.78	10.8			5.28
	ca.4.5	6.13	8.67			4.51
		5.89	7.61			4.40

(*Continued*)

Table 3.12.   (*Continued*)

	M6	M7	M8	M9	M10	M11
		5.22	6.11			3.10
		5.07	*ca.*4.2			2.77
		4.87				
		4.66				
		4.50				
		4.42				
		4.25				
		4.08				
		3.85				
		3.70				
		3.42				
		3.11				
		3.01				
		*ca.*4.4				
	9	7	38	p45	p54	p76
						$n = 16$
	RHO	$(2\text{-Et-}C_6)_8PcH_2$	$(C_{12}Pc)_2Lu$	$(C_{12}Salen)_2Ni$	$10(n = 80)$	$(C_6)_2DABCO\text{-}Br_2$
T in D_C		55	rt	200	122	120
	1061.416	1412.288	3893.33	633.57	380.53	722.86
	$Col_r(P2/a)$	$Col_{rd}(P2m)$	$Col_{ob.d}$	S_A	S_E	S_T

chain n-dodecyl groups. You can see five sharp peaks in the low angle region and two broad peaks in the wide angle region. These peaks are numbered for convenience from the low angle region as shown in the figure. Then, a vertical line is drawn from the top of these peaks, and the value of 2θ is accurately read. By substituting θ value from this 2θ into Equation (3.10), the corresponding interplanar spacing (spacing d_n ($n = 1$ to 7)) is calculated. In Equation (3.10) used for this calculation, λ is the wavelength of the X-ray source. In general, $\lambda = 1.5418\,\text{Å}$ of Cu Kα line is often used. Since the measurement limit is up to $2\theta = 2°$ in a general-purpose X-ray apparatus, the spacing can be measured up to $d = 44\,\text{Å}$ when the radiation source is Cu, but a spacing larger than the size cannot be measured. The measured spacing values are summarized in Table 3.12, M4. The broad peak of No. 6 ($d =$ about $4.7\,\text{Å}$) corresponds to the melting of the long chain alkyl groups in the periphery of $C_{12}PzCu$. The broad peak of No. 7 ($d =$ about $3.4\,\text{Å}$) corresponds to the fluctuation of the interdisk distance between disks. This corresponds to the reflection from the (001) plane of the 1D lattice. It should be noted that the X-ray reflection line appears as a broad peak when the inter-plane distance fluctuates, and appears as a sharp peak when the inter-plane distance does not fluctuate. From this broad peak, it can be determined that this liquid crystal phase is the disordered phase of the discotic liquid crystal. However, there is no standard for how sharp peaks can be judged as an ordered phases. Hence, it is nowadays said that it makes little sense at present to distinguish between disordered and ordered phases [9]. Since these two peaks No. 6 and No. 7 in the low angle region are not relevant to the calculation of the two-dimensional lattice, the analysis is considered with excluding them.

Next, what dimensionality this lattice showing by the peak Nos. 1 to 5 has will be considered sequentially from the first article of "golden rule for liquid crystal structure analysis" described above. For this purpose, the following estimation calculation is carried out.

$$\text{1D-Lamellar } 28.9\,\text{Å} \div 2 = 14.5\,\text{Å} \tag{3.25}$$

$$\text{2D-hexagonal } 28.9\,\text{Å} \div 3 = 16.7\,\text{Å} \tag{3.26}$$

$$\text{2D-tetragonal } 28.9\,\text{Å} \div \sqrt{2} = 20.4\,\text{Å} \tag{3.27}$$

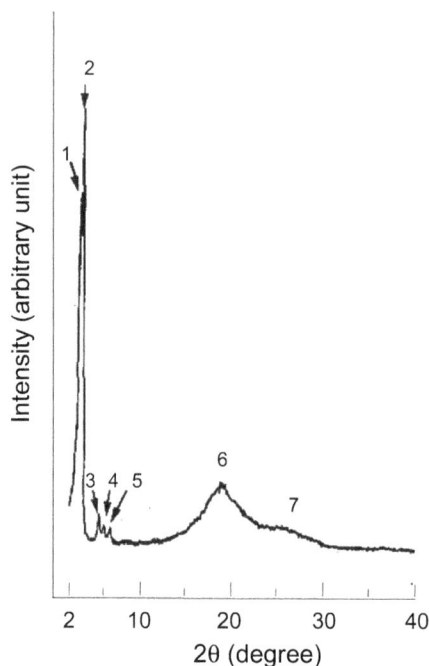

Figure 3.14.   X-ray diffraction pattern of $C_{12}PzCu$ at $150°C$ (M4 phase in Table 3.12).

See the observed values of M4 liquid crystal phase in Table 3.12. In consideration from Article ① and the Equation (3.25), the calculated value of $14.5\,\text{Å}$ agrees with the observed value of $14.7\,\text{Å}$ within the range of experimental error, whereas the observed value of $26.3\,\text{Å}$ and $16.4\,\text{Å}$ cannot be explained only from the ratio of lamella type. Next, considering from Article ② and Equation (3.26), the calculated value of $16.7\,\text{Å}$ may correspond to the observed value of $16.4\,\text{Å}$, but the observed value of $26.3\,\text{Å}$ cannot be explained so that it may not be a 2D-hexagonal lattice. Further, considering from the third article and Equation (3.27), the calculated value of $20.4\,\text{Å}$ does not correspond to any actually observed values at all, so that it is not a 2D-tetragonal lattice. Therefore, since it is neither 1D-lamella, nor 2D-hexagonal, nor 2D-tetragonal, it is considered to have a 2D-rectangular lattice or a 2D-oblique lattice. Looking at M4 in Table 3.12 again,

$$28.9\,\text{Å} : 14.7\,\text{Å} = 1 : 1/2$$
$$26.3\,\text{Å} : 13.2\,\text{Å} = 1 : 1/2$$

It looks like there are two lamella type ratios. This corresponds to Article ④, which is a feature of 2D-rectangular lattices.

From the above, it is inferred that this liquid crystal phase will be a Col_{rd} phase.

In a two-dimensional 2D-rectangular lattice, the strongest intensities are generally from the (20) and (11) planes. Hence, assuming $d_{20} = 28.9\,\text{Å}$ and $d_{11} = 26.3\,\text{Å}$, the lattice constants are calculated using Equation (3.2.1) and obtained as a $= 57.7\,\text{Å}$ and b $= 29.6\,\text{Å}$. At this time, we obey the rule that calculation should become $a > b$. (If not, the lattice constants are recalculated, assuming $d_{11} = 28.9\,\text{Å}$ and $d_{20} = 26.3\,\text{Å}$.) Then, draw a two-dimensional reciprocal lattice plane of $a^* = 1/a$ and $b^* = 1/b$ as shown in Fig. 3.15. The calculation of the lengths a^* and b^* in the reciprocal lattice is performed as follows:

$$a^* : \frac{1}{57.7} = 1.73 \times 10^{-2} \xrightarrow{\times 50} 0.867\,\text{cm}$$

$$b^* : \frac{1}{29.6} = 3.38 \times 10^{-2} \xrightarrow{\times 50} 1.69\,\text{cm}$$

(3.28)

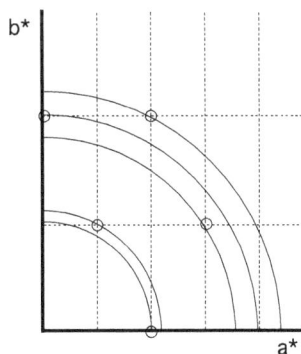

Figure 3.15. The reciprocal lattice of two-dimensional rectangular lattice for $C_{12}\text{PzCu}$, which corresponds to the X-ray diffraction pattern shown in Fig. 3.14.

Table 3.13.   Derivation method of the Debye–Scherrer Ring radius from the observed spacing values for M4 phase.

Peak No.	1	2	3	4	5
d_{obs} (Å)	28.9	26.3	16.4	14.7	13.2
$\frac{1}{d_{obs}} \times 10^{-2} = v$	3.46	3.80	6.10	6.80	7.58
$v \times 50 \rightarrow$ radius (cm)	1.73	1.90	3.05	3.04	3.78

A plane (a^*b^*) in 0.867 cm and 1.69 cm is drawn as the reciprocal lattice like as Fig. 3.15. Thus, the reciprocal lattice plane of a two-dimensional rectangular lattice is completed. Furthermore, in the reciprocal lattice plane are drawn the circles of reciprocal values of the spacings $(1/d)$. These circles correspond to the Debye-Scherrer rings. The radii are proportional to the reciprocals listed in Table 3.13. As shown in this table, the radius is obtained by multiplying the reciprocal by 50 and using cm as in Equation (3.28). You should draw these values as the quarter circles in Fig. 3.15. Indexing can be performed by reading the intersection of the reciprocal lattice and the quarter circle. Since (20) and (11) are assumed standards, it is natural that these quarter circles are placed on the intersection. The other three quarter circles are also in good agreement with the intersections (31), (02) and (22), so that it can be judged that the first assumption was correct. If these other three quarter circles do not coincide, recalculate again as $d_{11} = 28.9$ Å, $d_{20} = 26.3$ Å, etc., changing the previous assumption, and draw the reciprocal lattice plane and the quarter circles. It is important to repeat the calculation and drawing in this way in order to search for the smallest index (hk) values as possible to match all intersection points. As described above, if indexing is successful, it can be determined that this liquid crystal phase is the Col_{rd} phase.

Next, the symmetry of this two-dimensional rectangular lattice is determined to be $C2/m$ because $h + k = 2n + 1$ reflection lines do not appear from the extinction rule [10] in Fig. 3.13. For example, the reflection line from the (21) plane does not come out for this symmetry. Therefore, this liquid crystal phase has a symmetry of $C2/m$ and it can be identified as the $Col_{rd}(C2/m)$ phase in the four phases shown in Fig. 3.13. Here, for the first time, the phase structure can be specifically illustrated.

Furthermore, it is necessary to confirm whether this identification is correct by calculation of the number of molecules (Z) in the unit cell. Here, the stacking distance of the columnar liquid crystal phase is considered. Since the one-dimensional stacking distance in this column is in the c-axis direction, it corresponds to $d_{001} = c$ (lattice constant). In the case where the columnar liquid crystal phase is an ordered phase having a dimension of [2D$\oplus$1D], $d_{001} = c$ can be obtained immediately. However, in the disordered phases of Col_{hd}, $Col_{tet.d}$, and $Col_{ob.d}$, having only 2D dimensionality, d_{001} does not appear. Nevertheless, the van der Waals distance can be assumed to be $c = 3.3 \sim 3.5\,\text{Å}$ as the stacking distance for the organic disk-like molecule. In addition, in the case of Col_{rd} where the disk-kike molecules are inclined with respect to the column axis, the stacking distance can be assumed to be $c = 4 \sim 6\text{Å}$. This is only a guideline, so that it should be paid our attention that in some cases, Col_r may be inclined even for $c = 3.4\,\text{Å}$, and in some cases, it may be $c = 7\,\text{Å}$ without inclination when a Col_h is consisted of dimers.

Density $\rho(\text{g/cm}^3)$ of a liquid crystal phase is expressed by

$$\rho = (ZM)/(VN) \tag{3.29}$$

Here, Z is the number of molecules in a unit cell, M (g/mol) is a molecular weight, V (cm^3) is a unit cell volume, and N (number/mol) is an Avogadro's number. Therefore, the Z value is expressed as follows:

$$Z = (\rho VN)/M \tag{3.30}$$

It is very difficult to measure the density of the liquid crystal phase, but it is usually 0.8 to 1.2 (g/cm^3). In addition, the volume V of the unit cell in each liquid crystal phase of Col_{ho}, Col_{ro}, $Col_{tet.o}$ and $Col_{ob.o}$ can be expressed as one equation as

$$V = abc \sin \gamma, \tag{3.31}$$

where γ values are listed in Table 3.11.

The Z (the number of molecules in the unit cell) value must be an integer. If it does not become an integer, its identification will be false. If Z = 0.5 or 1.5, the phase identification is incorrect. In the case of this $C_{12}PzCu$, the Z value is calculated as follows:

$$Z = \frac{\begin{matrix} 1.0(g/cm^3) \times (57.7 \times 10^{-8}cm) \times (29.6 \times 10^{-8}cm) \\ \times (3.4 \times 10^{-8}cm) \times (6.02 \times 10^{23}/mol) \end{matrix}}{1930.58(g/mol)}$$

$$= 1.8 \approx 2 \tag{3.32}$$

This Z value is not perfect integer. It is attributable to $c = 3.4\,Å$ which represents only the average distance (Fig. 3.14) with very thermally fluctuates, and to the density assumed as 1. However, the obtained value $Z \approx 2$ is consistent, within the experimental error, with the theoretical value $Z = 2$ for $Colr(C2/m)$ shown in Fig. 3.13. Therefore, the above identification to C2/m symmetry is supported also by this Z value calculation.

3.5.3.2 $Col_{rd}(P2_1/a)$ phase

For the M5 phase [10] in Table 3.12, assuming $d_{11} = 32.0\,Å$ and $d_{20} = 28.3\,Å$, the calculation is carried out in the same way mentioned in the above section, and $a = 56.6\,Å$ and $b = 38.8\,Å$ are obtained. By drawing the reciprocal lattice and the quarter circles of the reverse values of spacings, the reflection lines at $18.6\,Å$, $12.6\,Å$ and $11.2\,Å$ can be indexed as (12), (13) and (42) respectively. From the extinction rules in Fig. 3.13, it can be concluded that this rectangular lattice has a symmetry of $P2_1/a$. As can be seen from Table 3.12, d_{001} does not appear for M5. So assuming $c = 5\,Å$, $Z = 1.97 \approx 2$ is obtained, which is consistent with the theoretical value $Z = 2$ for a rectangular lattice having a $P2_1/a$ symmetry. Therefore, this M5 phase can be identified as the Col_{rd} $(P2_1/a)$ phase.

3.5.3.3 $Col_{rd}(P2/a)$ phase

Looking at M6 phase [11] in Table 3.12, broad peaks at about $5.6\,Å$ and about $4.5\,Å$ correspond to fluctuation of the stacking distance and the melting of alkyl group, respectively. From the remained sharp peaks, the lattice symmetry is considered in the same manners as the above-mentioned rectangular mesophases. Assuming d_{11} 24.96 Å and d_{02} 17.47 Å, the lattice constants are calculated and $a = 35.7$ Å and $b = 34.9$ Å are obtained. In this case, it is assumed not d_{20} but d_{02} for accordance to the rule $a > b$. The reciprocal lattice and quarter circles are drawn for the indexation. The reflection lines at 16.00 Å, 9.76 Å and 8.98 Å can be indexed as (21), (23) and (40), respectively. When applied to the extinction rules in Fig. 3.13, the symmetry of this rectangular lattice matches both $P2_1/a$ and $P2/a$. Therefore, either cannot be specified from the extinction rules alone. However, when the number Z of molecules in a unit cell is calculated using $c = 5.6$ Å, Z $= 3.96 \approx 4$ can be obtained. From this Z value, the symmetry can be finally determined to be $P2/a$. Thus, this M6 phase is identified as the $\mathrm{Col}_{rd}(P2/a)$ phase. You can see from this example how important the Z value calculation is. Therefore, molecular weight is essential for the Z value calculation in liquid crystal structure analysis.

3.5.3.4 $Col_{rd}(P2m)$ phase

Looking at the M7 phase [12] in Table 3.12, a large number of reflection lines are obtained, and it may seem like a crystalline phase at first glance. However, a broad peak can be clearly observed at about 4.4 Å which corresponds to melting of the alkyl groups, so that it is not a crystal. Intentionally saying the conclusion at first, it is a Col_r phase having a $P2m$ symmetry shown in Fig. 3.13. There is no extinction rule in this symmetry, so that all reflection lines appear. Hence, a large number of reflection lines are observed for this M7 phase. Generally speaking, the number of reflection lines increases as the lattice symmetry decreases, and decreases as the symmetry increases. Since the reflections from a highly symmetrical lattice increase the probability that the wave phases are compensated with each other, many reflections become to disappear in higher symmetries. Therefore, in general, it is common knowledge that the number of reflection lines from liquid crystals is smaller than that of crystals, but it should be noted that there is such an exceptional example like M7.

Now, for the M7 phase in Table 3.12, assuming that $d_{10} = 20.4$ Å, $d_{01} = 18.6$ Å, the lattice constants $a = 20.4$ Å, $b = 18.6$ Å are obtained. By drawing the reciprocal lattice and the quarter circles, all the reflection lines are well matched with intersection points on this reciprocal lattice. Here, the two reflection lines having strongest intensity are assumed to d_{10} and d_{01} instead of d_{11} and d_{20}, unlikely as the previous rectangular symmetries. It is because there is no extinction rule in $P2m$ symmetry. As you can see from this example, it is indispensable for the X-ray liquid crystal structure analysis to understand the extinction rules. Next, assuming that the density is 1 and $c = 6$ Å, the number of molecules in the unit cell is calculated to be Z = $0.97 \approx 1$, which is consistent with the symmetry $P2m$. Thus, this M7 phase is identified as the $Col_{rd}(P2m)$ phase.

From the above analysis examples of M4 to M7 phases, it can be understood that the structural analysis of the liquid crystal phase needs to be performed in the following order.

① Estimation of n-dimensional lattice by "golden rule for liquid crystal structure analysis."
② Analysis by reciprocal lattice and indexing.
③ Application of the extinction rule.
④ Verification by Z value calculation.

3.5.4   *Liquid crystal phase having a two-dimensional hexagonal lattice (Col_h)*

When the X-ray data for the M1 phase[10] in Table 3.12 are checked sequentially from the first article of the "golden rule for liquid crystal structure analysis," the following estimation calculations are carried out, similarly to the previous Equations (3.25) to (3.27).

$$\text{1D-lamellar } 24.7 \text{ Å} \div 2 = 12.4 \text{ Å} \qquad (3.33)$$

$$\text{2D-hexagonal } 24.7 \text{ Å} \div \sqrt{3} = 14.3 \text{ Å} \qquad (3.34)$$

$$\text{2D-tetragonal } 24.7 \text{ Å} \div \sqrt{2} = 17.5 \text{ Å} \qquad (3.35)$$

Compare these results with the observed values of M1 phase in Table 3.12. In consideration of the first article and Equation (3.33), the calculated value of 12.4 Å agrees with the observed value of 12.3 Å within the experimental error, but the observed value of 14.2 Å cannot be explained only from the lamellar type ratio. Next, considering from the second article and Equation (3.34), the calculated value 14.3 Å corresponds to the observed value 14.2 Å. Further, considering from the third article and Equation (3.35), the calculated value 17.5 Å does not correspond to the observed value at all. Accordingly, this liquid crystal phase is estimated to be a 2D-hexagonal phase, Col_h.

Therefore, the observed value 24.7 Å is assumed as d_{10}, and the lattice constant is calculated to be $a = 28.5$ Å using Equation (3.4.1). By using the relationship of $a^* = \frac{2}{a\sqrt{3}}$ and $\gamma^* = 60°$ in Table 3.11, a two-dimensional hexagonal reciprocal lattice plane (a^* a^*; angle between two a^* axes is 60°) is drawn as Fig. 3.16. Furthermore, the circles (or quarter circles) having the radii of the reciprocals of spacing $(1/d)$ are drawn on the reciprocal lattice plane. Then, by reading the intersection points of the reciprocal lattice and the circles, the reflection lines at 24.7 Å, 14.7 Å, 12.3 Å, 9.28 Å can be indexed as (10), (11), (20), (21), respectively. There is no extinction rule for a two-dimensional hexagonal lattice, and all reflection lines appear. Therefore, we don't need here the consideration from extinction rule. Next, the Z value calculation is performed by using Equations (3.30) and (3.31). Assuming the density $\rho = 1$, the calculation is carried out using $c = 3.36$ Å, we can obtain $Z = 1.07 \approx 1$. This value is no

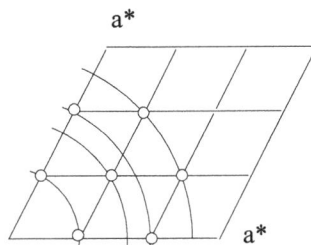

Figure 3.16. The reciprocal lattice of two-dimensional hexagonal lattice for Ni(12, $\bar{1}$) at 170°C (M1 phase in Table 3.12).

contradiction as Col_h. Therefore, this M1 phase is identified as the Col_{ho} phase.

For this Col_h phase, the analysis method using the ratio of Equation (3.5) may seem easier than the above reciprocal lattice method. However, in the case of the Col_h phase like as M2 phase [13] in Table 3.12, which shows a number of reflection lines, in the ratio method, the larger the h and k indies become, the more the candidate values become. In the many candidate values, it becomes very difficult to judge which one has the smallest error between the calculated value and the observed value, and the possibility of mistaking the indexation becomes exponentially higher. In that respect, the present analysis method using the reciprocal lattice is much better, because the error can be recognized at a glance as a distance from the intersection point on the graph, and it is unlikely to be misidentified.

3.5.5   *Liquid crystal phase having a two-dimensional tetragonal (square) lattice* (Col_{tet})

When the X-ray data for the M3 phase [12] in Table 3.12 are checked sequentially from the first article of the "golden rule for liquid crystal structure analysis," the following estimation calculations are carried out, similarly to the previous Equations (3.33) to (3.35).

$$\text{1D-lamellar } 38.8 \text{ Å} \div 2 = 19.4 \text{ Å} \tag{3.36}$$

$$\text{2D-hexagonal } 38.8 \text{ Å} \div \sqrt{3} = 22.4 \text{ Å} \tag{3.37}$$

$$\text{2D-tetragonal } 38.8 \text{ Å} \div \sqrt{2} = 27.4 \text{ Å} \tag{3.38}$$

When compared with the observed values of M3 phase in Table 3.12, each of the calculated values of (3.37) and (3.38) has an error of 1 to 2 Å, which makes judgment difficult. This is because the closer to $2\theta = 2°$, the greater the influence on the spacing d for the reading error of 2θ. When putting $\theta = 1.00$ and 1.10 into the Bragg's condition Equation (3.10), the error is 4.02 Å. On the other hand, when putting $\theta = 5.00$ and 5.10, the error is only 0.13 Å. From this simulation, you can see the big reading error in the very low angle region. In such a case, the estimation calculation is better to be performed based on $d_2 = 28.3\,\text{Å}$ or $d_3 = 20.0\,\text{Å}$, not based on the value $d_1 = 38.8\,\text{Å}$ in the lowest angle region. In this example, the third article of "golden rule for liquid crystal structure analysis" is applied on the basis of $d_3 = 20.0\,\text{Å}$. Since d_3 corresponds to the ratio $1/2$, it is calculated as $d_1 = 40.0\,\text{Å}$. Using this d_1 value, the estimation calculation is performed again.

$$\text{1D-lamella } 40.0 \text{ Å} \div 2 = 20.0 \text{ Å} \tag{3.39}$$

$$\text{2D-hexagonal } 40.0 \text{ Å} \div \sqrt{3} = 23.1 \text{ Å} \tag{3.40}$$

$$\text{2D-tetragonal } 40.0 \text{ Å} \div \sqrt{2} = 28.3 \text{ Å} \tag{3.41}$$

The calculated value $28.3\,\text{Å}$ of Equation (3.41) agrees with the observed value $28.4\,\text{Å}$ within the range of experimental error, and it is presumed that this liquid crystal phase is the Col_{tet} phase.

Therefore, assuming $40.0 \text{ Å} = d_{10}$, the lattice constant is calculated as $a = 40.0\,\text{Å}$ from Equation (3.6). By using the relationship of $a^* = \frac{1}{a}$ and $\gamma^* = 90°$ in Table 3.11, a two-dimensional tetragonal reciprocal lattice plane (a^*a^*; angle between two a^* axes is $90°$) is drawn as Fig. 3.17. Furthermore, the circles (or quarter circles) having the radii of the reciprocals of spacing $(1/d)$ are drawn on the reciprocal lattice plane. Then, by reading the intersection points of the reciprocal lattice and the circles, the reflection lines at $38.8\,\text{Å}$, $28.4\,\text{Å}$, $20.0\,\text{Å}$, $14.4\,\text{Å}$, $13.1\,\text{Å}$ can be indexed as (10), (11), (20), (20), (22), (30), respectively. There is no extinction rule for a two-dimensional tetragonal lattice, and all reflection lines appear. Therefore, we don't need here the consideration from extinction rule. Next, the Z value calculation is performed by using Equations (3.30) and (3.31). Assuming the density $\rho = 1$, the calculation is carried out using $c = 3.5\,\text{Å}$, we obtain $Z = 0.5$. In a Col_{tet} phase, the Z value must be 1. Since this compound $[(C_{18}\text{OPh})_8\text{Pc}]_2\text{Lu}$ is a sandwich complex of phthalocyanine, it was inappropriate that the stacking distance was assumed as $c = 3.5\,\text{Å}$. Accordingly, the stacking distance is assumed as $c = 7.0\,\text{Å}$ which is twice as usual, because the present compound is a sandwich complex [10]. The Z value is recalculated to

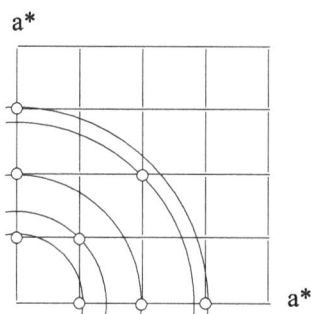

Figure 3.17. The reciprocal lattice of two-dimensional tetragonal lattice for $[(C_{18}OPh)_8Pc]_2Lu$ at rt (M3 phase in Table 3.12).

be just 1.0. This value is no contradiction as Col_{tet}. Therefore, this M3 phase is identified as the $Col_{tet.d}$ phase.

As described above, the phase structures of disk-like (discotic) liquid crystal can be analyzed by drawing a two-dimensional reciprocal lattice, using the spacings obtained from temperature-variable X-ray diffraction measurements.

I have not explained about M8 phase [14] in Table 3.12. The phase is a $Col_{ob.d}$ phase. In order to identify this oblique phase, we need the program for a computer. It is almost impossible to analyze without a computer. However, all the rest can basically be analyzed with paper, pencil, a ruler, a compass, and a calculator. If you will program the analysis software, the analysis time can be significantly reduced for all the liquid crystal phases. We have independently developed a software [15] that can analyze almost all the discotic and smectic liquid crystal phases. The computer system uses an old model NEC PC98 and a conventional language N88BASIC. The software is so convenient and useful that we have used it every day since 1990, although we think that this software has to be rewritten in Visual BASIC or C language suited to the times. Anyhow, see Ref. [14] for the identification of $Col_{ob.d}$ phase by using X-ray diffraction.

Thus, the author has described the X-ray liquid crystal structure analysis by using a novel **reciprocal lattice method**, taking the discotic liquid crystal phases as the examples. This **reciprocal lattice method** can be used not only for the discotic liquid crystal phases but also for the calamitic liquid crystal phases and lyotropic liquid crystal phases. So, in the next session, the author will analyze the structure of the calamitic liquid crystal phases using this **reciprocal lattice method**.

3.5.6 *Analysis of smectic phase*

3.5.6.1 *Smectic A phase*

For the M9 phase [16] in Table 3.12, calculating the ratio using only Article ① of the "Golden Rule for Liquid Crystal Structure Analysis,"

$$1D\text{-lamellar } 32.8 \text{ Å} \div 2 = 16.4 \text{ Å} \qquad (3.42)$$

$$1D\text{-lamellar } 32.8 \text{ Å} \div 3 = 10.9 \text{ Å} \qquad (3.43)$$

The calculated values of 16.4 Å and 10.9 Å agree with the observed values of 16.7 Å and 11.2 Å within experimental error.

$$32.8 : 16.7 : 11.2 = 1 : \frac{1}{2} : \frac{1}{3} \qquad (3.44)$$

It can be seen that it is a simple ratio specific to lamellar structures. Since no other reflection is observed, this phase can be estimated to be smectic A phase or C phase having only a 1D lamellar structure.

It cannot be determined any more from X-ray structure analysis which phase it has, so that the natural texture of this liquid crystal phase is observed with a polarization microscope. Since it exhibits a fan-shaped texture and a pseudo-isotropic region characteristic to smectic A phase [16], this liquid crystal phase is finally identified as a smectic A phase. Thus, X-ray structural analysis is not almighty for the identification of liquid crystal phases. Therefore, it is necessary to also observe the texture with a polarizing microscope at the same time, in order to make an overall judgment.

3.5.6.2 *Smectic E phase*

In order to estimate the dimensionality of the M10 phase [17] in Table 3.12 from "Golden rule for liquid crystal structure analysis," it is possible to calculate by multiplication inversely to the above Equations (3.42) to (3.43).

$$\text{1D-lamellar } 25.2 \text{ Å} \times 1 = 25.2 \text{ Å} \qquad (3.45)$$

$$\text{1D-lamella } 13.1 \text{ Å} \times 2 = 26.2 \text{ Å} \qquad (3.46)$$

Therefore, the observed value is in a ratio as

$$25.2 : 13.1 = 1 : \frac{1}{2} \qquad (3.47)$$

This simple ratio is characteristic to a lamella structure. No other reflection is observed in the low angle region, it can be estimated that this phase is a liquid crystal phase having a lamellar structure having an interlayer distance $c = 26.2$ Å. However, this liquid crystal phase shows three sharp peaks at 4.54 Å, 4.06 Å, and 3.26 Å in the higher angle region on a broad peak. Therefore, this phase has a 1D nature but it is not a smectic A phase or C phase having only a lamellar structure seen in Section 3.5.6.1. As already discussed in Fig. 3.8, when the compound is a calamitic liquid crystal, the 1D reflections appear only in the low angle region, while the 2D reflections appear only in the high angle region.

When the X-ray data of these 2D reflections in the high angle region are checked sequentially from the first article of the "golden rule for liquid crystal structure analysis," the following estimation calculations are carried out.

$$\text{1D-lamellar } 4.56 \text{ Å} \div 2 = 2.28 \text{ Å} \qquad (3.48)$$

$$\text{2D-hexagonal } 4.56 \text{ Å} \div \sqrt{3} = 2.63 \text{ Å} \qquad (3.49)$$

$$\text{2D-tetragonal } 4.56 \text{ Å} \div \sqrt{2} = 3.22 \text{ Å} \qquad (3.50)$$

See the observed values of M10 liquid crystal phase in Table 3.12. From the first article and Equation (3.48), the calculated value of 2.28 Å does not match the observed value. Next, considering from the second article and the equation (3.49), the calculated value of 2.63 Å does not match the observed value. Therefore, it does not have a 2D-hexagonal lattice. Furthermore, considering Article ③ and Equation (3.50), the calculated value of 3.22 Å agrees with the observed value of 3.26 Å within experimental error, whereas the observed value of 4.06 Å cannot be explained. Therefore, it does not have a 2D-tetragonal lattice. Thus, it is neither 1D-lamellar, nor 2D-hexagonal nor 2D-tetragonal, it can be considered to have a 2D-rectangular lattice or a 2D-oblique lattice. Since none of the smectic phases have a 2D-oblique lattice, it is presumed that this liquid crystal phase is a smectic E phase having a 2D-rectangular lattice (cf. Table 3.6).

It is usually expected that the strongest reflection intensity from the two-dimensional 2D-rectangular (rectangular) lattice is from the (20) and (11) planes. So, assuming $d_{11} = 4.54\,\text{Å}$ and $d_{20} = 4.06\,\text{Å}$, the lattice constants are calculated and obtained $a = 8.11\,\text{Å}$ and $b = 5.49\,\text{Å}$ using Equation (3.2.1).

By drawing the two-dimensional rectangular reciprocal lattice and the quarter circles of the reverse values of spacings in the same manner as in Fig. 3.15, the reflection lines at 4.54 Å, 4.06 Å, and 3.26 Å are well hit on the intersection points of (11), (20), and (21), respectively. From the extinction rules in Fig. 3.13, it can be concluded that this rectangular lattice has the symmetry of $P2_1/a$. Further, using the obtained lattice constants $a = 8.11\,\text{Å}$, $b = 5.49\,\text{Å}$, $c = 26.2\,\text{Å}$, and assuming the density is $1.1\,\text{g/cm}^3$, the Z value is calculated from Equation (3.30) to be Z = 2.0. This Z value supports that this phase has a smectic E structure having a symmetry $P2_1/a$ (cf. Fig. 3.10(3) structure). Thus, this M10 phase can be identified as the smectic E phase.

3.5.6.3 *Smectic T phase*

The estimation of the M11 phase [6] in Table 3.12 is performed from the "golden rule for liquid crystal structure analysis." As the result, a following ratio is obtained:

$$32.9 : 16.1 : 10.7 : 7.97 : 6.35 : 5.28 : 4.51 = 1 : \frac{1}{2} : \frac{1}{3} : \frac{1}{4} : \frac{1}{5} : \frac{1}{6} : \frac{1}{7}$$
$$(3.51)$$

The first seven observed values are simple ratios characteristic to a lamellar structure. Therefore, it can be estimated that this liquid crystal phase is a liquid crystal phase having a lamellar structure with an interlayer distance $c = 32.2$ Å.

Next, the remaining observed values of 4.40 Å, 3.11 Å, and 2.78 Å are checked from the first article of "golden rule for liquid crystal structure analysis". So, the following calculations are performed:

$$\text{1D-lamella } 4.40 \text{ Å} \div 2 = 2.20 \text{ Å} \tag{3.52}$$

$$\text{2D-hexagonal } 4.40 \text{ Å} \div \sqrt{3} = 2.54 \text{ Å} \tag{3.53}$$

$$\text{2D-tetragonal } 4.40 \text{ Å} \div \sqrt{2} = 3.11 \text{ Å} \tag{3.54}$$

See the observed values of M11 liquid crystal phase in Table 3.12. From the first and Equation (3.52), the calculated value of 2.20 Å does not match the observed value. Next, considering the second article and Equation (3.53), the calculated value of 2.54 Å does not match the observed value. Therefore, it does not have a 2D-hexagonal lattice. Further, considering from the third article and Equation (3.54), the calculated value of 3.11 Å just matches the observed value of 3.11 Å. Therefore, this smectic liquid crystal phase is considered to have a 2D-tetragonal lattice.

When assuming $4.40 \text{ Å} = d_{11}$, the lattice constant is calculated as $a = 6.22$ Å using Equation (3.6).

By drawing the two-dimensional tetragonal reciprocal lattice and the quarter circles of the reverse values of spacings in the same manner as in Fig. 3.17, the reflection lines at 4.40 Å, 3.11 Å, 2.78 Å are well hit on the intersection points of (11), (20), and (21), respectively. There is no extinction rule for two-dimensional tetragonal lattice. Therefore, we don't need here the consideration from extinction rule. Next, the Z value calculation is performed using Equations (3.30) and (3.31). Assuming the density $\rho = 1.0 \text{ g/cm}^3$, the Z value is calculated to be Z = 1.0. It consistently supports the smectic T phase. Thus, this M11 phase is identified as the smectic T phase.

As described above, the X-ray structural analysis of the liquid crystal phase using the **reciprocal lattice method** can be used not only for the discotic liquid crystal phases but also for the calamitic liquid crystal phases.

3.5.7 *Temperature-variable X-ray diffraction apparatus*

In order to measure the X-ray diffraction pattern of liquid crystal phases, an apparatus for heating the sample is required. Various apparatuses have been devised and made to date. Most methods employ a glass capillary used in the X-ray powder crystal method. In this method, a sample is packed in a capillary and heated. However, this capillary method has a problem that it is very difficult to pack a liquid crystal sample. In general, the liquid crystal material is extremely difficult to put in a thin capillary because it is not smooth loose fine powder but sticky mass. When the capillary is tapped on a desk to pack such a material, it breaks immediately. Therefore, we have developed and employed a novel method using a glass plate having a small hole illustrated in Fig. 3.18[A]. By using this special glass plate, it became extremely easy to pack such liquid crystal materials.

The setups of the temperature-variable sample holder and X-ray diffraction system we developed are shown in Figs. 3.18 and 3.19, respectively. Insert a glass plate (76 mm × 19 mm × 1.0 mm) with

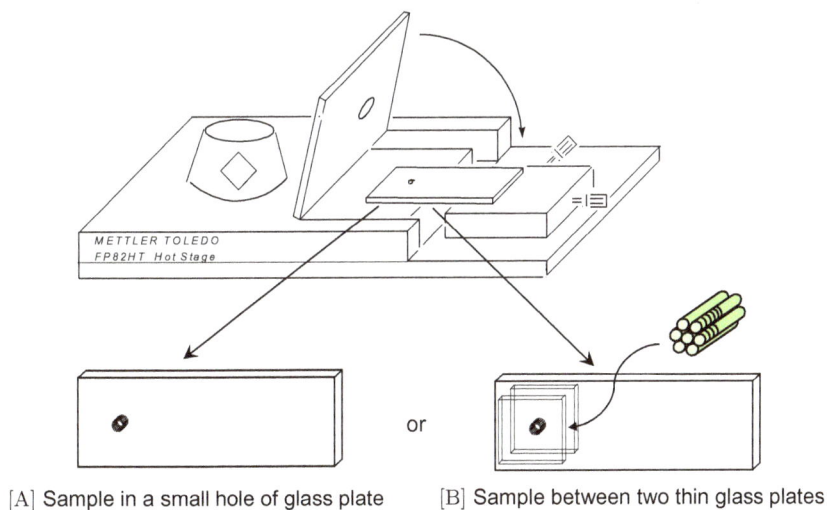

[A] Sample in a small hole of glass plate   [B] Sample between two thin glass plates

Figure 3.18. Setup of the temperature-variable sample holder.

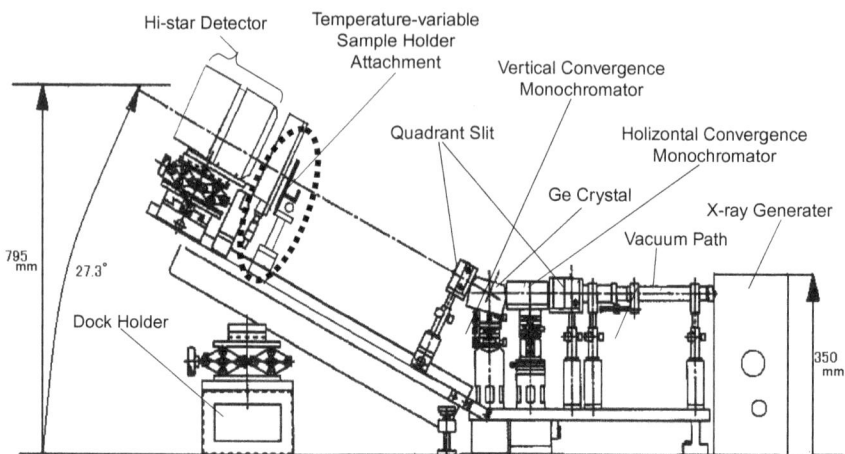

Figure 3.19. Setup of small-angle X-ray scattering (Bruker MAC SAXS) equipped with a temperature-variable sample holder.

holes (diameter 1.5 mm) into the temperature-variable sample holder of the Mettler FP 82HT hot stage as shown in Fig. 3.18. There are two ways to set the sample as shown in Figs. 3.18[A] and [B]. In the case of a non-oriented sample, the method [A] is used. At this time, the hole can be filled with a powder sample (about 1 mg) [18]. When used this glass plate, the powder sample can be filled in the hole by strongly pressing with a spatula. This sample glass plate is usable for all condensed phases including fluidic nematic phases and isotropic liquids, as it does not flow out of the pores due to surface tension. In the case of an alignment sample such as homeotropic alignment, method [B] can also be used (cf. Section 3.5.8).

As can be seen in Fig. 3.19, the generated X-rays are bent by two convergence monochromators to produce a point X-ray beam (diameter = 1.0 mm). The point beam passes through a hole in the variable temperature sample holder. The measurable range is 3.0 Å to 110 Å, and the temperature range is room temperature to 375°C. All liquid crystal phases can be identified by using a small angle X-ray diffractometer (Bruker Mac SAXS system) equipped with our temperature-variable sample holder, which employs a Mettler FP82HT hot stage [18].

3.5.8 *X-ray structure analysis method using orthogonal vector complement subspaces*

As described in Section 3.3.3, the [2D⊕1D]-dimensional space is not a three-dimensional space but a direct sum of a two-dimensional subspace and a one-dimensional complement subspace. The two-dimensional subspace and the one-dimensional subspace are vector spaces orthogonal to each other. An advanced X-ray structural analysis method using this concept of orthogonal vector space is described below.

The author and his co-workers noticed in 2009 that many liquid crystalline phthalocyanine-fullerene (Pc-C_{60}) dyads show a very interesting X-ray diffraction peak H [19]. Figure 3.20 shows temperature-dependent small angle X-ray diffraction patterns of $(12, 8)$PcCu-C_{60} (**18b**) as an example. As can be seen from these diffraction patterns, the complex **18b** additionally shows a very large

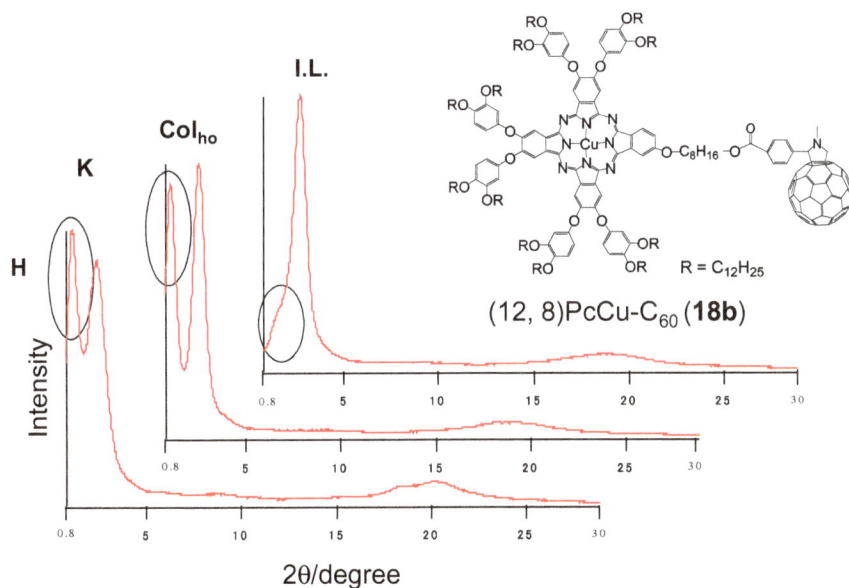

Figure 3.20. Temperature-dependent small angle X-ray diffraction patterns of $(12, 8)$PcCu-C_{60} (**18b**). During our study, we noticed a very interesting XRD peak H for all the Pc-C_{60} Dyads. The peak H could not be assigned as a reflection from any 2D-lattices known up to date.

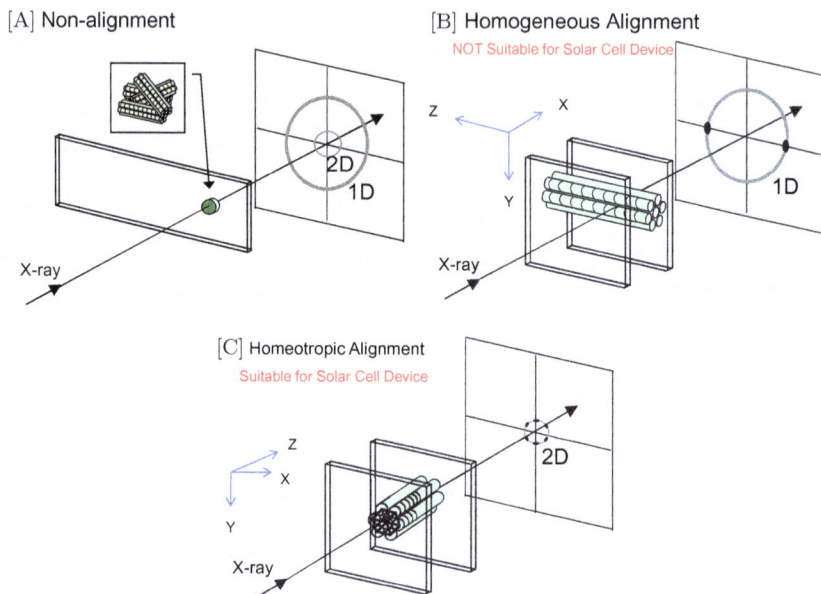

Figure 3.21. For clarification of the origin of Peak H by using three different alignments of the Col_{ho} mesophase: [A] non-aligned sample in a hole, [B] homogeneously aligned sample and [C] homeotropically aligned sample between two glass plates.

peak (designated as H) in the crystal phase K and the liquid crystal phase Col_{ho} in the very low angle region. Also in the isotropic liquid I. L. it appears as a small shoulder at the same position. The peak H could not be assigned as a reflection from any 2D-lattices known till that time.

Therefore, the authors considered to clarify the origin of peak H by using three different orientation samples shown in Fig. 3.21. As shown in this figure, there are three types of [A] non-aligned sample in a hole, [B] homogeneously aligned sample and [C] homeotropically aligned sample between two glass plates.

The reflections obtained from these samples are as follows:

- In the method [A], the Col_{ho} liquid crystal phase packed in small holes is a non-aligned polydomain, so that we can observe the X-ray reflections from both 2D and 1D lattices as shown in Fig. 3.21[A].

The small and large rings correspond to reflections from the 2D lattice (in the XY plane) and the 1D lattice (in the Z-axis direction), respectively.

- In the method [B], the perfect homogeneously oriented Col_{ho} liquid crystal phase between the two glass plates gives the reflection from only the 1D lattice in the Z-axis direction as shown in Fig. 3.21[B].
- In the method [C], the perfect homeotropically aligned Col_{ho} liquid crystal phase between two glass plates gives the reflection from only the 2D lattice in the XY plane as shown in Fig. 3.21[C].

The different appearances of X-ray diffraction in Figs. 3.21[B] and [C] are resulted from in a vector space divided into two orthogonal subspaces. In the viewpoint of linear algebra, the two-dimensional subspace and the one-dimensional subspace divided in a Col_{ho} liquid crystal phase are **orthogonal** to each other. In this vector space, if you stand on the 2D subspace, you cannot see the 1D subspace. Oppositely, if you stand on the 1D subspace, you cannot see the 2D subspace. It is so-called "**Parallel World**" in SF novels. In the perfectly aligned sample of Col_{ho} phase, we can actually see this phenomenon.

If peak H in Fig. 3.20 is a periodicity along the Z-axis direction of the column, peak H should disappear in the homeotropic alignment sample, as shown in Fig. 3.21[C].

Therefore, both the non-aligned sample and the homeotropic aligned sample of the dyad **18b** were prepared, and temperature-variable small-angle X-ray scattring (SAXS) patterns were measured [19 h, i, k ∼ m].

In Fig. 3.22, solid black and red lines indicate the SAXS patterns of the non-aligned sample and the homeotropically aligned sample of the dyad **18b** at 95°C, respectively. The dotted curve shows the SAXS pattern of two cover glass plates without sample.

As can be seen from these patterns, in the non-aligned sample, both the peak H and the (100) reflection peak are clearly observed. On the other hand, in the homeotropic aligned sample, the (100) reflection peak was observed but the peak H completely disappeared.

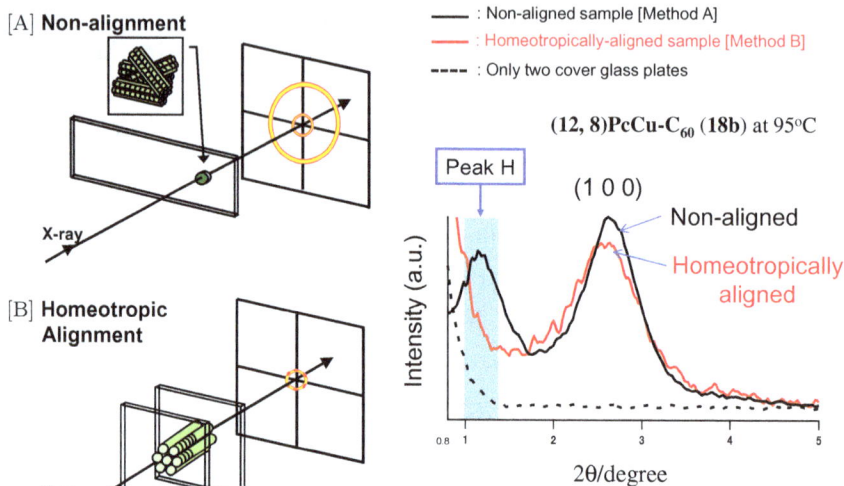

Figure 3.22. The **black** and **red** solid curves show XRD patterns for [A] the non-aligned sample and [B] the **homeotropically aligned** sample of Col_{ho} mesophase between two glass plates.

This means that the peak H has periodicity along the column in Z-axis direction.

From the further detailed examination, in the Col_{ho} liquid crystal phase of this Pc-C_{60} dyad **18b**, it was established that the fullerenes (C_{60}) takes a helical structure around the column formed by phthalocyanines (Pc), and that this peak H corresponds to the pitch of the helix of the C_{60} moieties [19h, i, k ∼ m]. From precise analysis of the SAXS data mentioned above and additional CD spectra, the helical structure of fullerenes in the Col_{ho} mesophase was established. Fullerenes pile up in left-handed and right-handed helicity in a ratio of 50:50 as illustrated in Fig. 3.23. This helical structure resembles spiranthes flower as can be seen from this figure. Therefore, this unique structure was named "**Spiranthes-like Supramolecular Structure**."

As can be seen from the above example of Col_{ho} phase, the two-dimensional subspace and the one-dimensional subspace are orthogonal to each other in a vector space.

Spiranthes-like Supramolecular Structure

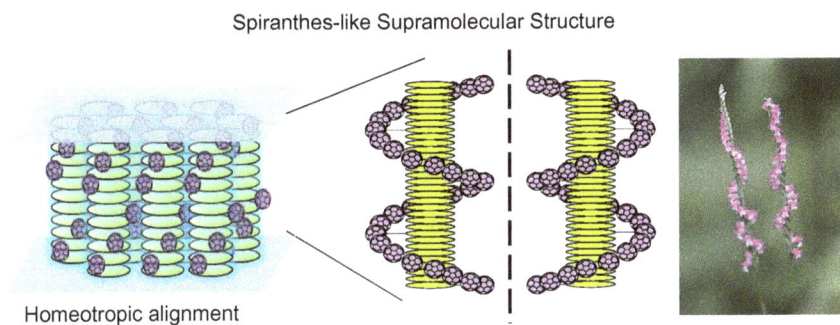

Homeotropic alignment

Figure 3.23. From precise analysis of the SAXS data mentioned above and additional CD spectra, the helical structure of fullerenes in the Col_{ho} mesophase was established. Fullerenes pile up in left-handed and right-handed helicity in a ratio of 50:50. This helical structure resembles spiranthes flower. Hence, this supramolecular structure is named as Spiranthes-like Supramolecular Structure.

By using these orthogonal vector subspaces, it is possible to clarify in which direction the unknown peak is periodic. For this purpose, it is essential to prepare a perfect homeotropic alignment sample or a perfect homogeneous alignment sample. Since 2010, Imabori and his collaborators have reported that a similar liquid crystalline Pc-C_{60} dyad has a helical structure of C_{60} [20]. However, they did not describe at all how the perfect homeotropic alignment sample was prepared. Furthermore, their identified rectangular columnar (Col_r) phase cannot be in principle homeotropically aligned. Only the Col_h and Col_{tet} phases that are face-to-face stacking can be homeotropically aligned [19m]. Therefore, they cannot in principle hold the claim that the helical peak disappears in the aligned sample. Furthermore, they did not write any alignment sample preparation methods, so that nobody can perform the follow-up experiment. Therefore, their proof of the helical structure is inadequate and needs to be re-experimented. The ambiguities of their proof may be due to their poor understanding of the principle of this orthogonal vector subspaces.

As can be seen from the above two examples, this advanced X-ray structural analysis method uses the fact that a two-dimensional subspace and a one-dimensional subspace are orthogonal to each other in one vector space. In this case, it is essential to describe in detail how the perfect alignment sample is prepared and to prove from a polarization microscope that the perfect alignment sample is surely obtained. Readers are particularly required to pay attention on this point when using this analysis [19m].

3.6 Miscellaneous Discussion

3.6.1 $3D < [2D \oplus 1D]$ *phases*

Although it has not been mentioned until now because it becomes complicated, there are a number of liquid crystal phases having dimensionality of $[2D \oplus 1D]$ with leaving a little bit of 3D dimensionality. It is denoted as $3D < [2D \oplus 1D]$ in Table 3.6. There has been an endless debate as to whether these phases are crystalline or liquid crystalline, and no conclusions have been drawn until now. The 3D dimensionality can be evaluated by the size of the Miller index appeared in the X-ray structure analysis. The liquid crystal phases of $3D < [2D \oplus 1D]$ show only very small Miller indies of $h + k + l \approx 5$ $(h, k, l \leq 2 \sim 3)$, whereas ordinary crystals show very big Miller indies of $h + k + l = 20 \sim 50$.

Figuratively speaking, when you stand on one molecule in such a phase and look at your surroundings, two or three around you in the top, bottom, left, right, front, back are certainly three-dimensionally arranged, but over four around you their position of the molecules is not clear by the thermal fluctuation. Such a state is the $3D < [2D \oplus 1D]$ phase. On the other hand, when you stand on one molecule in a normal crystal and you look at your surroundings, even distant molecules are clearly seen because the molecules are arranged infinitely in three dimensions. This is the definition of the crystal itself. So, how many molecules are aligned in the crystal phase and liquid crystal phase, respectively? What numbers of molecules is the boundary between crystal and liquid crystal? It seems to be correct that the present $3D < [2D \oplus 1D]$ phases are judged as crystal phases because they have three-dimensionality at least, but they do not line up in long distance three-dimensionally as in ordinary crystal phases. Therefore, it is difficult to judge these $3D < [2D \oplus 1D]$ phases as the perfect crystal phases. They are the phases in a gray zone, and we cannot judge from X-ray structural analysis whether they are the soft crystal phases or the hard liquid crystal phases. However, very interestingly, when creating a binary mixture phase diagram, these 3D < [2D⊕1D] phases lie above two freezing point depression curves and a eutectic point, and below a liquidus curve of the isotropic liquid. A freezing point depression curve is the boundary between crystal and liquid [21]. Thus, it is apparent that the 3D < [2D⊕1D] phase has the liquid nature thermodynamically.

Therefore, it is our present conclusion that we can say that it is a liquid crystal phase from a thermodynamic point of view.

3.6.2 *3D mesophases*

Another problem is whether plastic crystals, smectic D phase, Cub phase, etc. are crystal phases or mesophases (liquid crystal phases). Each of them shows three-dimensionality, so that it can be judged to be a crystal phase from X-ray structure analysis. However, when creating a binary mixture phase diagram, these phases lie above two freezing point depression curves and a eutectic point, and below a liquidus curve of the isotropic liquid. Thus, it is apparent that the these phases have the liquid nature thermodynamically [22]. It can be judged from the thermodynamic point of view that they are liquid crystal phases (mesophases).

3.6.3 *Steel (austenite phase)*

These three-dimensional mesophases (plastic crystals, smectic D phase, Cub phase, etc.) and 3D < [2D⊕1D] phases in the gray zone show not spontaneous fluidity but viscosity or plasticity. When pressed, they are not broken unlikely as crystals. Here we consider a binary system of iron and carbon. Castings containing more than 4.2% of carbon are brittle and fragile when pressed down. On the other hand, steels containing less than 4.2% of carbon have malleability without cracking even if pressed. Very interestingly, the steel of the austenitic phase is above a eutectoid point very similar to eutectic point in the Fe-C binary phase diagram [23]. The phase above such a (pseudo) eutectic point is liquid-like and soft. In that sense, the austenite phase can be said to be a mesophase.

As mentioned above, the author expressed his current opinion on the three open issues from thermodynamic point of view. These are the issues that should be taken up by future researchers when reviewing the definition of mesophases between crystal and liquid. The author would be very happy if his present idea would contribute to the development of mesophase research in the future.

References

[1] Shuzo Seki and Hiroshi Seki, Glassy states of pure compounds-thermodynamical study, Kagaku Sosetsu, No. 5: *Non-equilibrium State and Relaxation Process*, Tokyo Daigaku Syuppankai, Tokyo, 1974, pp. 225–256 (in Japanese).

[2] Flying-seed-like liquid crystals: Y. Takagi, K. Ohta, S. Shimosugi, T. Fujii and E. Itoh, *J. Mater. Chem.*, **22**, 14418–14425 (2012).

[3] Database of liquid crystalline compounds for personal computer: *LiqCryst Version 5*, V. Vill, LCI Publisher and Fujitsu Kyushu Systems Limited, 2010.

[4] K. Ohta, *Mol. Cryst. Liq. Cryst.*, **658**, 13–31 (2017).

[5] S_L, S_G, S_H, S_J and S_K are rigid liquid crystal phases having a little bit three dimensionality: see Table I in G. W Gray, and J. W. G Goodby, *Smectic Liquid Crystals-Textures and Structures*, Leonard Hill, Grasgow and London, 1984.

[6] K. Ohta, T. Sugiyama and T. Nogami, *J. Mater Chem.*, **10**, 613–616 (2000).

[7] Toshio Sakurai, *Introduction to X-ray Crystallographic Analysis*, Shokabo, 1983 (in Japanese).

[8] K. Ohta, T. Watanabe, H. Hasebe, Y. Morizumi, T. Fujimoto, I. Yamamoto, D. Lelièvre and J. Simon, *Mol. Cryst. Liq. Cryst.*, **196**, 13–26 (1991).

[9] K. Ohta, Y. Inagaki-Oka, H. Hasebe and I. Yamamoto, *Polyhedoron*, **19**, 267–274 (2000).

[10] K. Ohta, S. Azumane, T. Watanabe, S. Tsukada and I. Yamamoto, *Appl. Organometallic Chem.*, **10**, 623–635 (1996).

[11] (a) J. Billard, J. C. Dubois, C. Vaucher and A. M. Levelut, *Mol. Cryst. Liq. Cryst.*, **66**, 115–122 (1981); (b) C. Destrade, P. Foucher,

H. Gasparoux, Nguyen Huu Tinh, A. M. Levelut and J. Malthete, *Mol. Cryst. Liq. Cryst.*, **106**, 121–146 (1984).

[12] T. Komatsu, K. Ohta, T. Watanabe, H. Ikemoto, T. Fujimoto and I. Yamamoto, *J. Mater. Chem.*, **4**, 537–540 (1994).

[13] K. Hatsusaka, K. Ohta, I. Yamamoto and H. Shirai, *J. Mater. Chem.*, **11**, 423–433 (2001).

[14] T. Komatsu, K. Ohta, T. Fujimoto and I. Yamamoto, *J. Mater. Chem.*, **4**, 533–536 (1994).

[15] Shinshu University Ohta Laboratory Homemade Software *"Bunseki kun"* Ver.1.1; Masahiro Ando, Master Thesis, Shinshu University, Appendix, p. A1–A120, 1993.

[16] K. Ohta, Y. Morizumi, T. Fujimoto, I. Yamamoto, K. Miyamura and Y. Gohshi, *Mol. Cryst. Liq. Cryst.*, **214**, 161–169 (1992).

[17] K. Ohta, H. Akimoto, T. Fujimoto and I. Yamamoto, *J. Mater. Chem.*, **4**, 61–69 (1994).

[18] Y. Kanai, H. Akimoto and K. Ohta, *Mol. Cryst. Liq. Cryst.*, **648**, 130–147 (2017).

[19] (a) L. Tauchi, M. Shimizu and K. Ohta, *The Proceedings of the Annual Conference on Liquid Crystal Society*, 2009-09-13 ~ 15, 1C02; (b) K. Ota (=Ohta), *Jpn. Kokai Tokkyo Koho*, JP 2011-132180 (A)-2011-07-07 (Priority number: JP2009-0293501; Submission Date: 2009-12-24) (Patent); (c) H.-T. Nguyen-Tran, L. Tauchi, T. Kamei and K. Ohta. *The Proceedings of the 90th CSJ Annual Meeting*, 2010-03-26 ~ 29, 3 E4-06; (d) L. Tauchi, M. Shimizu, T. Fujii, H.-D. Nguyen-Tran, T. Kamei and K. Ohta, *23rd International Liquid Crystal Conference*, 2010-07-11 ~ 16, Krakow Poland, P-1.106; (e) K. Ohta, L. Tauchi, H.-T. Nguyen-Tran, M. Shimizu, T. Kamei and T. Kato, *The Proceedings of Pacifichem 2010*, Hawaii, USA, 2010-12-15~19, MATL-255 (invited lecture); (f) L. Tauchi, M. Shimizu, K. Ohta and E. Itoh, *The Proceedings of Pacifichem 2010*, Hawaii, USA, 2010-12-15 ~ 19, MATL-795; (g) M. Shimizu, L. Tauchi, T. Nakagaki, A. Ishikawa, E. Itoh and K. Ohta, *J. Porphyrins Phthalocyanines*, **17**, 264–282 (2013); (h) L. Tauchi, T. Nakagaki, M. Shimizu, E. Itoh, M. Yasutake and K. Ohta, *J. Porphyrins Phthalocyanines*, **17**, 1080–1093 (2013); (i) A. Ishikawa, K. Ono, K. Ohta, M. Yasutake, M. Ichikawa and E. Itoh, *J. Porphyrins Phthalocyanines*, **18**, 366–379 (2014); (j) M. Yoshioka, K. Ohta, Y. Miwa, S. Kutsumizu and M. Yasutake, *J. Porphyrins Phthalocyanines*, **18**, 856–868 (2014); (k) A. Watarai, S. Yajima, A. Ishikawa, K. Ono, M. Yasutake and K. Ohta, *ECS Transactions*, **66**, 21–43 (2015); (l) A. Watarai, K. Ohta and M. Yasutake, *J. Porphyrins Phthalocyanines*, **20**, 1444–1456 (2016); (m) K. Ishikawa, A. Watarai,

M. Yasutake and K. Ohta, *J. Porphyrins Phthalocyanines*, **22**, 693–715 (2018).

[20] (a) W. Nihashi, H. Hayashi, Y. Shimizu, T. Umeyama, Y. Matano and H. Imahori, *The Proceedings of the 39th Symposium on the Fullerens, Nanotubes and Graphene Research Society*, 2010-09-07; 3P-2; (b) W. Nihashi, H. Hayashi, T. Umeyama, Y. Matano and H. Imahori, *The Proceedings of the 91st CSJ Annual Meeting*, 2011-03-26 ∼ 29; 2F2-13; (c) H. Hayashi, W. Nihashi, T. Umeyama, Y. Matano, S. Seki*, Y. Shimizu* and H. Imahori*, *J. Am. Chem. Soc.*, **133**, 10736–10739 (2011); (d) H. Hayashi, W. Nihashi, T. Umeyama, Y. Matano, S. Seki, Y. Shimizu and H. Imahori, *J. Amer. Chem. Soc.*, 2017, **139**, 13957–13957. Correction to Ref. 20(c).

[21] L. Richter, D. Demus and H. Sackmann, *J. Phys. Colloques*, **37**, C3-41–C3-49 (1976).

[22] N. Lindner, M. Kollbel, C. Sauer, S. Diele, J. Jokiranta and C. Tschierske, *J. Phys. Chem. B*, **102**, 5261–5273 (1998).

[23] (a) J. B. Austin, *Metal Handbook*, p. 1181, American Society for Metals, Cleaveland, 1948; (b) See Fig. 7.24 in p.272 of the textbook W. J. Moore, *Physical Chemistry*, 4th ed., Japanese Translation by Ryoichi Fujishiro, 1974, Tokyo Kagaku Dojin Tokyo.

Chapter 3. Exercises

1. Describe the characteristics of liquid crystal and plastic crystal, respectively, and explain the difference.

2. Classify the types of liquid crystals and explain them briefly.

3. Using equations of (1) 3D orthogonal system, (2) 3D hexagonal system, (3) 3D tetragonal system, and (4) 3D monoclinic system, explain logically each case of the stepwise disintegration of crystal lattice by heating or solvent addition (See Section 3.3.1) and Table 3.3)

4. Give several examples of the liquid crystalline compounds showing the 1D, 2D, [2D ⊕ 1D] [1D ⊕ 1D] and [1D ⊕ 1D ⊕ 1D] mesophases. Be sure to draw each of the molecular structures and indicate each of the source original papers.

5. Calculate the extinction rules of a two-dimensional rectangular lattice for (1) $C2/m$, (2) $P2_1/a$, (3) $P2/a$ symmetry, from these liquid crystal structure factors.

6. Problems for calculation of the extinction rule when a new symmetry of two-dimensional rectangular lattice appears:

(6.1) $D_{L \cdot rec}(P2_1 2_1)$ and $D_{L \cdot rec}(P2_1 1)$ phases: K. Ohta, R. Higashi, M. Ikejima, I. Yamamoto and N. Kobayashi, *J. Mater. Chem.*, 1998, **8**, 1979–1991. Read this paper, and calculate the extinction rule when the symmetry of the two-dimensional lattice is $(P2_1 2_1)$ and $(P2_1 1)$.

(6.2) $M(Pa2_1)$ phase: Y. Abe, K. Nakabayashi, N. Matsukawa, H. Takashima, M. Iida, T. Tanase, M. Sugibayashi, H. Mukai, and K. Ohta, *Inorg. Chim. Acta,* 2006, **359**, 3934–3946. Read this paper and calculate the extinction of $(Pa2_1)$. This symmetry was a new two-dimensional rectangular lattice which had never been found till that time.

7. When a disk-like molecule $C_{12}PzH_2$ was heated to 150°C, the X-ray diffraction lines were obtained as follows:

Spacing d(Å)	intensity	line width
27.6	Large	Sharp
15.9	Medium	Sharp
13.3	Small	Sharp
ca. 4.7	Medium	Broad
ca. 3.5	Medium	Broad

Identify this liquid crystal phase by using "Reciprocal Lattice Method." Also calculate the lattice constants.

8. (a) $C_{12}PzCu$ could be prepared by metallation of the metal-free derivative $C_{12}PzH_2$ in Q7. When this Cu(II) complex was heated to 150°C, the X-ray diffraction lines were obtained as follows:

Spacing d(Å)	intensity	line width
28.9	Large	Sharp
26.3	Large	Sharp
16.4	Medium	Sharp
14.7	Small	Sharp
13.2	Small	Sharp
ca. 4.7	Medium	Broad
ca. 3.4	Medium	Broad

Identify this liquid crystal phase by using "Reciprocal Lattice Method." Also calculate the lattice constants.

(b) Considering the extinction rules for 2D lattices, determine the lattice symmetry of this mesophase.

9. The attached sheets are a list of X-ray diffraction data for discotic liquid crystal phases. Choose 10 favorite data from this list and identify each liquid crystal phase. Be sure to identify all the columnar phases by using the reciprocal lattice method other than a simple discotic lamellar phase. Give the Miller index for each diffraction line and calculate the lattice constants. If necessary, determine the symmetry of the lattice.

XRD data of discotic liquid crystals.

(1)Yamaguchi1	(5)Yamaguchi5	7.46	12.6	24.5
Mw = 4172.74	Mw = 2690.38	6.83	11.2	22.0
36.0	37.0	6.50	ca.4.8	18.5
21.4	34.8	5.94	(10)Azumane2	12.8
9.55	18.8	5.64	Mw = 3685.15	ca.4.8
ca.4.3	13.4	5.34	31.1	(13)Azumane5
(2)Yamaguchi2	10.7	ca.4.5	27.2	Mw = 4134.01
Mw = 4172.74	ca.4.4	(8)Oka1	19.8	34.7
34.4	(6)Yamaguchi6	Mw = 1280.22	11.8	30.5
19.9	Mw = 2690.38	21.6	ca.4.8	17.7
ca.9.1	41.7	12.6	(11)Azumane3	15.8
ca.4.4	36.3	10.9	Mw = 3909.62	13.6
(3)Yamaguchi3	22.0	8.20	36.0	9.94
Mw = 4234.21	10.5	7.25	31.5	8.83
36.3	ca.4.6	6.28	19.7	7.95
21.2	(7)Ikejima1	6.01	17.8	ca.4.8
ca.9.3	Mw = 2148.16	5.41	13.1	(14)Azumane6
ca.4.4	28.0	ca.4.5	9.65	Mw = 4134.01
(4)Yamaguchi4	16.74	3.38	8.58	34.8
Mw = 2241.54	14.69	(9)Azumane1	ca.4.8	30.1
33.2	11.2	Mw = 3685.15	(12)Azumane4	24.7
29.8	9.87	32.0	Mw = 3909.62	21.9
16.6	8.54	28.3	33.4	18.6
ca.4.4	8.22	18.6	29.1	14.9

(*Continued*)

(*Continued*)

13.0	8.30	(25)Azumane16	6.94	16.2
11.5	7.27	Mw = 4137.11	ca. 4.4	14.0
ca.4.8	6.78	35.4	ca. 3.9	10.6
(15)Azumane7	4.96	29.6	(30)Ito2	ca. 4.6
Mw = 2667.36	ca.4.8	19.3	Mw = 1637.24	(37)Nishizawa3
31.3	4.75	18.1	26.0	Mw = 3508.53
23.0	3.82	16.3	15.0	22.7
11.2	(21)Azumane12	ca.4.8	13.0	13.2
10.4	Mw = 4141.95	(26)Azumane17	9.84	11.4
7.41	30.1	Mw = 3693.09	ca. 4.3	8.64
ca.4.8	16.9	34.8	(31)Ito3	ca. 4.7
3.77	15.1	27.5	Mw = 1662.45	(38)Aoki1
(16)Hatsusaka11	11.2	18.2	25.1	Mw = 3941.86
Mw = 4710.87	9.93	15.0	14.7	31.8
31.9	8.78	13.8	12.7	25.4
22.7	6.98	11.4	9.64	17.4
9.09	ca.4.8	9.16	ca. 4.4	15.2
ca.4.6	(22)Azumane13	8.07	(32)Ito4	12.9
(17)Azumane8	Mw = 4141.95	7.01	Mw = 2111.31	8.57
Mw = 2667.36	35.8	ca.4.8	29.0	7.23
34.0	30.1	(27)Nakai4	16.9	ca. 4.4
27.4	19.6	Mw = 2756.16	14.6	(39)Nakai1
17.1	16.7	41.4	11.2	Mw = 1087.63
13.8	15.0	21.1	ca. 4.6	26.5
12.1	12.3	14.2	(33)Ito5	13.7
ca.4.8	9.82	ca. 8.6	Mw = 4101.29	9.21
(18)Azumane9	8.75	ca. 4.4	29.8	ca. 6.5
Mw = 2662.52	7.49	ca. 3.7	17.2	ca. 4.8
32.0	ca.4.8	(28)Azumane18	14.9	(40)Nakai2
22.6	(23)Azumane14	Mw = 3693.09	ca. 4.5	Mw = 1199.85
10.2	Mw = 4141.95	34.4	(34)Ito6	30.9
7.59	36.2	27.5	Mw = 4101.29	15.2
7.14	30.0	17.5	31.2	10.2
ca.4.8	19.3	14.9	17.7	ca. 6.5
ca.3.7	14.9	13.8	15.7	ca. 4.8
(19)Azumane10	12.3	11.4	11.5	(41)Nakai3
Mw = 2662.52	10.8	9.90	ca.4.6	Mw = 1312.06
32.9	9.82	9.26	(35)Nishizawa1	33.1
26.6	8.75	8.15	Mw = 1730.33	16.9
13.5	7.43	7.06	23.8	11.2
10.9	ca.4.8	ca.4.8	13.9	ca. 6.5
ca.4.8	(24)Azumane15	(29)Ito1	11.9	ca. 4.7
(20)Azumane11	Mw = 4137.11	Mw = 1637.24	9.20	(42)Nakai5
Mw = 2442.93	31.5	24.4	ca. 4.6	Mw = 3417.01
29.7	18.4	19.8	(36)Nishizawa2	28.8
21.4	8.72	14.6	Mw = 2179.19	18.1
9.39	ca.4.8	10.2	27.9	13.1

(*Continued*)

10.4	13.3	15.0	ca. 9.6	8.98
9.32	10.4	13.0	ca. 4.6	7.09
7.95	ca. 7.9	9.86	(59)Ban10	5.92
6.92	ca. 4.2	ca. 7.2	Mw = 2117.66	5.16
6.17	ca. 3.6	ca. 4.7	28.1	ca. 4.6
ca. 4.2	(49)Ban2	ca. 3.5	16.2	ca. 3.5
(43)Nakai6	Mw = 3537.09	(55)Ban4	ca. 6.3	(65)Ban16
Mw = 3865.87	33.6	Mw = 3873.73	ca. 4.6	Mw = 3508.53
31.1	29.5	38.2	ca. 3.6	22.7
23.2	22.3	30.4	(60)Ban11	13.2
18.7	17.2	19.4	Mw = 2566.52	11.4
14.2	11.5	16.4	31.2	8.64
13.1	10.7	15.1	18.1	ca. 6.2
11.3	8.75	14.2	15.6	ca. 4.7
10.6	ca. 7.2	12.9	ca. 6.4	ca. 3.6
8.84	ca. 4.2	10.9	ca. 4.7	(66)Ban30
8.07	ca. 3.7	10.1	ca. 3.5	Mw = 4832.11
6.40	(50)Ban3	8.95	(61)Ban12	29.7
ca. 4.2	Mw = 3873.73	8.42	Mw = 1730.33	16.9
(44)Hatada1	45.8	7.76	24.3	15.2
Mw = 1126.57	22.8	7.00	14.0	10.2
24.8	15.4	ca. 4.3	12.1	ca.7.7
12.4	10.7	ca. 3.6	9.21	ca.4.6
8.32	9.25	(56)Ban5	8.20	(67)Ban17
6.23	ca. 6.6	Mw = 4210.35	6.70	Mw = 3957.39
ca.4.4	ca. 4.2	24.9	ca. 4.6	26.6
(45)Hatada2	3.6	17.0	ca. 3.5	21.0
Mw = 1182.68	(51)Ban8	12.7	(62)Hatsusaka8	15.6
27.9	Mw = 6998.80	10.2	Mw = 4262.01	13.6
14.0	33.2	8.62	32.0	10.3
ca.4.3	20.0	7.38	21.8	ca.7.6
(46)Hatada3	ca. 9.0	6.31	12.2	ca.4.3
Mw = 1238.79	ca. 4.6	4.23	ca.4.6	3.28
28.8	(52)Ban9	3.54	(63)Ban14	(68)Ban18
14.5	Mw = 8345.39	(57)Ban6	Mw = 2179.19	Mw = 3957.39
9.60	38.8	Mw = 4210.35	27.9	25.1
7.23	22.2	40.0	16.2	14.5
ca.4.3	ca. 9.2	30.6	14.0	12.6
(47)Hatada4	4.16	20.5	10.6	9.51
Mw = 1294.89	(53)Ban7	13.9	ca. 6.5	8.47
32.5	Mw = 6998.80	10.6	ca. 4.6	ca.6.7
16.4	35.6	ca. 7.1	ca. 3.6	ca.4.7
10.1	20.4	ca. 4.3	(64)Ban15	(69)Ban19
ca.4.4	ca. 9.4	ca. 3.6	Mw = 2628.05	Mw = 4406.25
(48)Ban1	ca. 4.4	(58)Ban9b	31.0	29.1
Mw = 3537.09	(54)Ban13	Mw = 8345.39	18.0	22.5
38.5	Mw = 1954.76	37.6	15.6	16.8
20.3	26.3	21.9	11.9	14.7

(*Continued*)

(*Continued*)

11.2	*ca.* 7.3	15.9	12.9	*ca.*4.7
9.88	6.70	13.7	9.74	(87)Ban38
8.26	*ca.* 4.7	10.5	8.48	Mw = 5736.79
7.42	*ca.* 3.6	9.23	7.31	33.1
*ca.*6.5	(75)Ban25	7.52	6.42	19.2
*ca.*4.4	Mw = 3934.39	*ca.*7.2	*ca.*4.7	16.6
3.28	27.2	*ca.*4.7	(83)Ban34	12.7
(70)Ban20	20.8	(79)Ban29	Mw = 4390.21	9.61
Mw = 4406.25	15.9	Mw = 4832.11	29.1	*ca.*7.0
26.6	13.7	29.8	22.2	6.41
15.5	12.2	17.1	17.1	*ca.*4.5
13.4	10.5	15.0	14.7	*ca.*3.5
10.2	9.21	11.1	11.3	(88)Hatsusaka1
7.73	8.04	8.08	9.82	Mw = 3588.72
*ca.*7.0	7.74	*ca.*7.6	8.28	31.5
*ca.*4.7	6.90	5.39	6.50	11.9
(71)Ban21	*ca.* 4.6	*ca.*4.8	*ca.* 4.4	8.78
Mw = 4855.11	4.28	*ca.*3.5	3.29	6.91
27.8	*ca.* 3.3	(80)Ban31	(84)Ban35	*ca.*4.5
16.5	(76)Ban26	Mw = 3492.49	Mw = 4390.21	(89)Hatsusaka5
14.2	Mw = 3934.39	24.1	27.2	Mw = 4262.01
10.8	25.5	14.0	15.9	35.5
9.52	14.9	12.4	13.7	20.5
ca. 7.8	12.9	9.07	9.17	13.4
ca. 4.7	9.82	7.94	*ca.*7.4	10.2
(72)Ban22	8.67	*ca.* 7.4	5.62	*ca.*4.5
Mw = 5303.97	7.23	6.55	*ca.*4.7	(90)Hatsusaka2
30.4	6.08	5.43	3.91	Mw = 3588.72
17.8	*ca.*4.7	*ca.* 4.7	(85)Ban36	33.1
15.4	(77)Ban27	(81)Ban32	Mw = 4839.07	29.4
*ca.*7.1	Mw = 4383.25	Mw = 3941.35	29.0	24.1
*ca.*4.7	28.8	27.0	16.8	13.5
(73)Ban23	22.0	20.7	14.3	12.4
Mw = 5752.83	17.0	15.7	11.0	11.3
33.5	14.8	13.7	9.57	10.1
19.2	11.2	10.4	8.13	9.07
16.5	9.85	9.27	*ca.*7.6	8.22
12.3	9.45	7.62	*ca.*4.7	7.87
*ca.*7.0	8.25	*ca.* 6.5	(86)Ban37	6.48
*ca.*4.7	7.74	6.01	Mw = 5287.93	4.98
(74)Ban24	7.43	5.53	29.9	*ca.*4.5
Mw = 3485.53	5.65	*ca.* 4.5	17.7	(91)Hatsusaka3
24.5	*ca.* 4.5	3.27	15.3	Mw = 3588.72
14.0	3.30	(82)Ban33	11.7	33.1
12.1	(78)Ban28	Mw = 3941.35	8.50	29.1
9.17	Mw = 4383.25	25.1	*ca.*7.4	13.2
7.92	27.0	14.7	6.81	11.2

(*Continued*)

10.2	21.9	19.8	8.50	16.7
9.64	12.3	17.3	*ca.*4.6	12.6
8.15	9.05	13.2	(103)Ariyoshi6	9.07
7.32	*ca.*4.6	11.5	Mw = 3907.44	*ca.*4.6
6.40	(95)Hatsusaka9	9.56	35.9	(107)Ariyoshi11
5.92	Mw = 4710.87	8.60	21.0	Mw = 3624.85
*ca.*5.0	37.4	7.87	18.2	30.4
*ca.*4.5	21.6	*ca.*4.6	13.8	21.8
(92)Hatsusaka4	18.7	(99)Ariyoshi7	12.3	*ca.*4.7
Mw = 3588.72	14.2	Mw = 3907.44	10.2	*ca.*3.6
30.2	10.7	32.3	9.24	(108)Ariyoshi12
26.6	9.26	23.0	8.37	Mw = 3961.49
20.0	8.51	*ca.*4.7	*ca.*4.6	37.4
11.5	*ca.*4.6	*ca.*3.5	(104)Ariyoshi8	21.6
8.52	(96)Hatsusaka10	(100)Ariyoshi3	Mw = 3624.85	18.9
*ca.*4.6	Mw = 4710.87	Mw = 3570.80	34.8	14.2
(93)Hatsusaka6	38.9	32.9	20.2	10.4
Mw = 4262.01	36.0	19.1	17.5	8.68
36.8	21.4	16.6	13.3	*ca.*4.5
33.7	18.1	12.6	11.8	(109)Ariyoshi13
27.5	15.3	11.1	9.86	Mw = 3961.49
20.0	11.6	9.21	8.07	36.8
14.6	9.24	*ca.*4.6	*ca.*4.5	21.4
13.6	*ca.*4.6	(101)Ariyoshi4	(105)Ariyoshi9	18.5
12.9	(97)Ariyoshi1	Mw = 3570.80	Mw = 3624.85	14.1
11.0	Mw = 3570.80	30.8	34.2	12.3
9.96	34.8	22.3	19.8	10.7
8.86	20.2	*ca.*4.7	17.4	10.2
8.58	17.7	*ca.*3.6	13.3	9.32
7.87	13.3	(102)Ariyoshi5	9.63	*ca.*4.6
7.36	9.82	Mw = 3907.44	8.72	(110)Ariyoshi14
4.98	8.77	36.2	8.02	Mw = 3961.49
*ca.*4.5	8.01	21.1	*ca.*4.6	32.6
(94)Hatsusaka7	*ca.*4.5	18.4	(106)Ariyoshi10	23.1
Mw = 4262.01	(98)Ariyoshi2	13.9	Mw = 3624.85	*ca.*4.6
32.3	Mw = 3570.80	10.3	32.7	*ca.*3.5
29.9	34.2	9.24	19.2	

Below the (No.) sample name, Mw means the molecular weight and the observed spacings (Å).

General References

Chapter 1

Barlow Physics and Chemistry (8th edition); Kittel *Introduction to Solid Physics* Chapters 1–2

Toshio Sakurai, *Guide to X-ray Crystallography*, Shokabo, Tokyo, 2003.

Chapter 2

WC McCrone, Polymorphism, in *Physics and Chemistry of the Organic Solid State*, Vol. II, D. Fox *et al.* (Ed.), John Wiely & Sons, Int., New York 1965.

Chapter

Kazuchika Ohta The history of liquid crystal, *Chemistry* 49, 848–849 1994.

Kusabayashi, Takenaka Structure and physical properties of liquid crystal *Journal of Organic Synthesis Association* 42(2), 1984.

Kazuchika Ohta Identification of discotic liquid crystal phase by X-ray structural analysis, *Journal of the Liquid Crystal Society of Japan: Liquid Crystal* 6, 61–71 2002.

Kazuchika Ohta, Dimensionality and Hierarchy of Liquid Crystal Phase: X-ray Structure Analysis of the Dimensional Assembly. http://hdl.handle.net/10091/17016, Shinshu University, 2013.

Index